Rainfed Agriculture and Watershed Management

Sohan Singh Walia
School of Organic Farming
Punjab Agricultural University
Ludhiana, Punjab, India

Karmjeet Kaur
School of Organic Farming
Punjab Agricultural University
Ludhiana, Punjab, India

Tamanpreet Kaur
School of Organic Farming
Punjab Agricultural University
Ludhiana, Punjab, India

ISBN 978-981-99-8424-4 ISBN 978-981-99-8425-1 (eBook)
https://doi.org/10.1007/978-981-99-8425-1

This Springer imprint is published by the registered company Springer Nature Singapore Pte Ltd.
The registered company address is: 152 Beach Road, #21-01/04 Gateway East, Singapore 189721, Singapore

Preface

Feeding teeming global human population, especially in the developing world, remains a huge challenge for agricultural scientists and farmers alike. It must however be emphasized that in the context of overall world food production, rainfed agriculture has played and would continue to play a crucial role in the global food production as a large part of agriculture is rainfed (~80%) and contributes over 50% to the global food supply. Moreover, rainfed areas are also where poorest of the poor live, and these areas remain hotspots of poverty and malnutrition. Even a modest increase in productivity of the vast rainfed areas would result not only in a large increase in global food volume but would also stabilize these areas with enhanced food availability and rural development. Most certainly, agricultural development is at the heart of sustainable development of dryland areas. Increasing the productivity of rainfed areas, especially in the semi-arid and arid regions of the world with impoverished soil resource base and water shortages is the greatest challenge of the twenty-first century.

Water shortage is recognized as the primary constraint in rainfed areas and remains a deterrent to the use of external inputs of plant nutrients even at a modest level although the soil resource base in these areas is fragile and marginal compared to their irrigated counterpart. Recent research demonstrates that the soils in rainfed areas are not only thirsty (water shortage) but also hungry (nutrient deficiencies). Nevertheless, recent focus on rainfed agriculture along with concomitant enhanced investment in rainfed agriculture, a prerequisite to sustainably increase the productivity in rainfed areas, has helped to intensify research in these areas and develop technology packages that simultaneously address both water shortage and nutrient disorders that are holding back the potential of rainfed areas. These science-based interventions have provided highly encouraging results from extensive on-farm evaluation of the improved technology or through the upgradation of rainfed agriculture.

This book discusses various issues related to Rainfed Agriculture. The topics covered are most relevant in view of growing interests in rainfed agriculture technologies. The focus is on new concepts and approaches in dryland and rainfed lands; rainfed farming—introduction, history, characteristics, distribution, and problems; soil and climatic conditions prevalent in rainfed areas; efficient management of rainfed crops; water harvesting and moisture conservation; study of mulches and antitranspirants; concept of watershed resource management; factors affecting watershed management and impact of watershed management program on sustainable agriculture; drought and its management; stress physiology; plant ideotype, their types and ideotype for dryland farming; soil and water conservation techniques.

Ludhiana, Punjab, India Sohan Singh Walia
Ludhiana, Punjab, India Karmjeet Kaur
Ludhiana, Punjab, India Tamanpreet Kaur

Contents

About the Authors

Sohan Singh Walia is working as Director in the School of Organic Farming, Punjab Agricultural University, Ludhiana. He started organic farming research as a pioneer work at PAU, Ludhiana as Ph.D. student (2001–2004) and then worked on organic farming and integrated farming systems in All India Co-ordinated Research Project on Integrated Farming Systems and All India Network Project on Organic Farming. Dr. Walia has to his credit more than 450 research and extension publications, 9 books, 10 teaching manuals, 7 extension folders, and 22 book chapters. He has handled 24 research projects and is currently handling 5 research projects. The chapter on organic farming and integrated farming system in package of practices was included during 2004–2005 and 2017–2018, respectively. Sixty eight recommendations have been included in the package of practices for mass adoption by the Punjab farmers, especially resource conservative cropping systems, nine organic farming-based cropping systems, production technology of cultivation of direct seeded rice. In addition, he was involved in technologies regarding application of consortium in sugarcane, turmeric, potato, onion, maize, wheat crops; integrated nutrient management in maize/soybean, rice residue management for the mass scale adoption under Punjab conditions. Developed Integrated Farming System Research Model comprising dairy, fishery, horticulture, vegetables, agroforestry, and vermi-composting. He has been involved in teaching 93 courses. He has guided 4 Ph.D. and 11 M.Sc. students as major advisor. Dr. Walia has organized 20 training programs, delivered 62 invited lectures, 350 training lectures, and 52 TV and radio talks. He has attended 10 international and 75 national conferences/seminars/workshops. He was appreciated for outstanding work on integrated (2007–2017) and organic farming during QRT review (2012–2017). Recipient of Best Organic Farming Centre Award (2019) by ICAR; Dr. M. S. Randhawa Best Book Award (2017); Fellow, Indian Ecological Society (2016); Gold medal from the Society of Recent Development in Agriculture at International Conference (2013); ISA Best Paper Award along with cash prize of Rs. 5000/- from the Indian Society of Agronomy (2011) for paper entitled, “Alternate cropping systems to rice-wheat for Punjab”; ISA P. S. Deshmukh Young Agronomist Award (2005) by the Indian Society of Agronomy; Fellow, Society of Environmental Sciences (2004). The AICRP on Integrated Farming System, Ludhiana centre received the Best Centre Award (2014) from ICAR-the Indian Institute of Farming Systems Research,

Modipuram, Meerut. Received Dr. Gurbaksh Singh Gill Gold Medal, Merit Certificate for M.Sc. (Agronomy); S.S. Labh Singh Gold Medal and Merit Certificate for B.Sc. Agri. (Hons.). Dr. Walia has acquired advance training in rice production systems from the School of Agriculture, Food and Wine, University of Adelaide, South Australia.

Karmjeet Kaur is working as Senior Research Fellow (2019–till date) under project entitled "All India Co-ordinated Research Project on Integrated Farming Systems" at the School of Organic Farming, Punjab Agricultural University, Ludhiana. She holds M. Sc. in Agriculture (Agronomy) from Khalsa College, Amritsar.

Tamanpreet Kaur is working as Senior Research Fellow (2018–till date) under project entitled "All India Co-ordinated Research Project on Integrated Farming Systems" at the School of Organic Farming, Punjab Agricultural University, Ludhiana. She holds M. Sc. in Agriculture from Punjab Agricultural University, Ludhiana. She has 8 research papers, 13 extension articles, 1 booklet on integrated farming systems and 3 book chapters.

Definition, Concept, and Characteristics of Dry Land Farming

1

Abstract

Dry land farming is the cultivation of crops in the regions with annual rainfall more than 750 mm and crop growing season of 75–120 days with shortage of moisture availability. Single crop/multi-cropping systems are followed in these areas. Dry spells and crop failure are less frequent. However, wind erosion/water erosion are the main constraints of dry land areas. In these regions, maximum quantity of water is conserved by soil and water management practices. Dry land agriculture occupies 60% of cultivated area and supports 40% of human population and 60% livestock population and thus plays a distinct role in Indian agriculture. Rainfed area constitutes more than 90% of the area under sorghum, groundnut, and pulses, 82 to 85% under maize and chickpea, 78% under cotton, 65.8% under rapeseed/mustard. Moreover, 61.7, 44.0, and 35.0% area under rice, barley, and wheat, respectively, is rainfed.

Keywords

Cropping systems · Dry land farming · Erosion · Rainfall · Rainfed

1.1 Definition of Dryland Farming

An improved system of cultivation by which the maximum quantity of water is conserved by water and soil management is known as dry land farming.

In dry land farming, crops are cultivated in the areas having more than 750 mm of annual rainfall. Failure of crops is relatively less frequent in dryland farming. Growing period of crops varies between 75 and 120 days. In dry land farming, for successful crop production, practices for conservation of moisture are essential. However, for vertisols or black soils adequate drainage is required.

S. S. Walia et al., *Rainfed Agriculture and Watershed Management*,
https://doi.org/10.1007/978-981-99-8425-1_1

In India, nearly 75% of the total arable land that is about 108 million hectares out of the total 143 million hectares of arable land are under rain fed area. Due to its dependence on the intensity and frequency of rainfall in rainfed areas, production of crops becomes relatively difficult. In such areas, production of crops is known as rainfed farming because of no possibility of any irrigation, and even of life-saving irrigation.

In these areas, rainfall distribution is not proper, erratic, and uncertain and annual rainfall range is between 400–1000 mm. In certain areas, there is not more than 500 mm of total annual rainfall. Depending upon the rainfall amount, the production of crops is called dry land farming and regions under this farming are known as dry lands. In India, out of 108 million hectares of total rainfed area, about 47 million hectares are under dry lands. In India, dry lands contribute 42% in the total production of food grain. About 75% of pulses are produced from these areas and arid and semi-arid regions produce greater than 90% of millet, pulses, sorghum, and peanut. Therefore, rainfed farming and dry land play a crucial role in agricultural production. Dry lands are water deficient and characteristics include high day temperature during summer, evaporation rates, soil erosion, and runoff and low humidity. Saline and less fertile soils are often found in such areas. Period of scarcities and famines occurs due to complete or partial failure of crops because of inadequacy and uncertainty of rainfall.

1.2 Features of Dry Land Areas

- Limited, uncertain, and ill-distributed annual rainfall.
- Drought, flood, etc. occur extensively.
- Surface of soil is undulating.
- Large holdings occurrence.
- Extensive agriculture, i.e., monocropping is practiced.
- Fields are relatively large in size.
- Similar types of crops are grown by all the farmers of a particular region.
- Crop yield is very low.
- The produce market facility is poor.
- Farmer's economy is poor.
- Farmers and cattle with poor health.

1.3 Depending on the Amount of Rainfall Received, Dryland Agriculture Can Be Divided into Three Categories

Dry farming: It is cultivation of crops in regions with annual rainfall less than 750 mm. Crop failure is most common due to prolonged dry spells during the crop period. These are arid regions with a growing season (period of adequate soil moisture) less than 75 days. Moisture availability is acute and single crop/intercropping systems are followed. Dry spells are most common and wind erosion is the main constraint in dry farming. Moisture conservation practices are necessary for crop production.

Dryland farming: Cultivation of crops in regions with annual rainfall more than 750 mm. In spite of prolonged dry spells, crop failure is relatively less frequent. These are semi-arid tracts with a growing period between 75 and 120 days. Moisture conservation practices are necessary for crop production. Moisture availability is shortage and single crop/multi-cropping systems are followed. Dry spells are less frequent and wind erosion/water erosion are the main constraints in dryland farming. However, adequate drainage is required especially for vertisols or black soils.

Rainfed farming: Cultivation of crops in regions with annual rainfall more than 1150 mm. These are humid regions with growing period more than 120 days. Crops are not subjected to soil moisture stress during the crop period. Moisture availability is enough and intercropping/double cropping systems are followed. Water erosion is the main constraint in rainfed farming. Emphasis is often on disposal of excess water.

1.4 Importance and Dimensions of Dry Land Farming in Indian Agriculture

1. In dry farming areas, nearly 70% population of rural parts depends on crops success or failure for livelihood.
2. In Indian agriculture, dry land agriculture plays an important role by occupying cultivated area about 60% and supporting 40% of human and 60% livestock population.
3. The contribution of rainfed agriculture to the total food grain is 42%, oilseeds 75%, pulses 90%, and cotton production 70%.
4. Rainfed area constitutes greater than 90% of the area under sorghum, groundnut, and pulses, 82 to 85% under maize and chickpea, 78% under cotton, 65.8% under rapeseed/mustard. Moreover, rice, barley, and wheat constitutes 61.7, 44.0, and 35.0% area as rainfed.
5. Cereal grain production from 3 ha of dryland is equivalent to production from 1 ha irrigated crop.
6. Due to problems related to raising water table, salinity build up and nutrient exhaustion in irrigated agriculture, existing resources conservation is necessary and more production from drylands would be challenged. Therefore, new strategies are required to make the fragile dryland ecosystems more sustainable and productive.
7. The systems that supply food as well as fuel to the people and fodder to the cattle can be practiced in drylands. For example, agroforestry, social forestry, and horti-silvi-pasture. For ecological maintenance, it forms a suitable vegetative cover.

Dimensions of the problem: In India, dry farming districts are in majority which constitutes about 60% of the total cultivated area. Crops like millets, pulses, oilseeds, cotton, etc. are grown on most of this area. Areas under dry farming in the country are Karnataka, Tamil Nadu, Maharashtra, Madhya Pradesh, Andhra Pradesh, Rajasthan, Gujarat, Uttar Pradesh, Punjab, and Haryana. The Deccan plateau in south India is rain shadow area consisting of Karnataka parts (Raichur, Bellary, Dharwad, Tumkur, Belgaum, Kolar, Gulbarga) and Maharashtra (Pune, Parbhani, Aurangabad, Sholapur). In Andhra Pradesh, the dry farming areas are found in Kurnool, Anantapur, Kadapa, Mahbubnagar, Chittoor, and Nalgonda districts.

(a) In India, 60% of total cultivable area under dryland agriculture.
(b) Out of total geographical area more than 30% area with below 750 mm rainfall.
(c) Low rainfall area constitutes 84 districts in India.
(d) Providing irrigation takes a long time as well as expensive to all the drylands.
(e) Even after providing all the irrigation potential, 55% area remains as rainfed in India.

2 Soil and Climatic Parameters with Special Emphasis on Rainfall Characteristics

Abstract

Almost all the regions of the India constitutes dry land and the major soil types in these dry regions are red soils (Alfisols), black soils (Inceptisols and Vertisols), laterite soils (Ultisols), and desert soils (Aridisols). Moreover, in undulating lands of sub-Himalayan regions under coniferous forest sub-montane soils occur. Furthermore, the average annual precipitation is deficit in relation to evaporation in a dry climate which is of two types, arid and semi-arid climate. Arid climate is an extreme dry climate where the average annual precipitation is less than 500 mm, and the amount of precipitation is insufficient for crop production. However, in semi-arid climate, the average annual precipitation is more than 500 mm, and crop production is possible with dry farming methods or with supplemental irrigation. Among the different climatic parameters, rainfall is an important factor influencing the crop production in dry regions due to rainfall variability, intensity and distribution of rainfall, variations or aberrations in monsoon behavior like late onset of monsoon, early withdrawal of monsoon and prolonged dry spells.

Keywords

Annual precipitation · Crop production · Dry climate · Evaporation · Soil types

2.1 Dry Land Soils

Dry lands are found in all the regions of the India and red soils, black soils, laterite soils, and desert soils are the major soil types in these dry regions. In arid and semi-arid regions of the country where the annual temperature is high and rate of evaporation generally exceeds rainfall, saline and sodic soils are found.

S. S. Walia et al., *Rainfed Agriculture and Watershed Management*, https://doi.org/10.1007/978-981-99-8425-1_2

Red Soils (Alfisols)

In India and especially in dry land, red soil is an important soil group. These soils are generally red or reddish brown or yellowish brown in color because of ferric oxides coating on soil particles. Red earth with loose topsoil and red loams are morphological types.

Friable structure, light texture, absence-free calcium carbonates, and lime concretions and soluble salts contents are main features of red soils. These soils are medium in cation exchange capacities, slightly acidic to slightly alkaline in nature and near-base saturated. Kaolinite with an admixture of illite is dominant clay mineral. Permeability of red soils is moderate, and these soils are well drained. Surface crust formation, excessive gravel, and erosion susceptibility to high slopes are some of the problems in these soils. Nitrogen and phosphorus nutrients as well as organic matter content are low in these soils. In these soils, the micronutrient-like zinc is occasionally found in low amount. For raising crops, potassium is generally in adequate amount and range of soil pH is 6–7.

Black Soils (Vertisol/Inceptisol)

Another important group of soils in dry land conditions is black soils. The clay content of these soils is high and soil reaction is neutral to slightly alkaline. The color of soils is dark grey to black and develops deep cracks in summer.

Black soils have low permeability and impeded drainage. It founds in monsoon climate areas, mostly in semi-arid and sub-humid type. Higher swelling and shrinkage properties of these soils are imparted by higher amount of clay and montmorillonite type of clay mineral.

Base nutrients like calcium, magnesium, and potassium are found higher in these soils. However, organic matter content, available nitrogen and phosphorus and sometimes zinc are in low amount. The range of soil pH is 7.5–8.5. The soil texture is highly clayey and the clay content of soil between 35–60% and sometimes up to 80%.

High stickiness, swelling, plasticity, and shrinkage are the characteristics of black soils. The reaction of black soils is neutral to slightly alkaline, cation exchange capacity is high, high base status and calcareous. The moisture holding capacity of soil is also high, runoff is severe and is prone to erosion. Entisols and Inceptisols order constitute shallow black soils, while the Vertisols constitute deep medium black soils.

2.2 Laterite Soils (Ultisols)

These soils have good drainage conditions and hydraulic conductivity. Under conditions of heavy rainfall, silica is liberated and leached downwards, and the upper strata of the soil become rich in iron and aluminum oxides. Laterization is the action in question. High-level and low-level laterites are two categories of laterites. The laterites at the top are thin and gravely; therefore, not useful for agriculture and are found capping the peaks of hills and plateaus in the high areas. Low-level laterites are of considerable agricultural importance and comprising of clay and loam in coastal areas on both sides of the peninsula. Rainfall that is modest in intensity increases lateralization.

In top layers, a laterite soil typically has a loam texture, and the depth of this layer varies and is naturally eroded. Laterite soils are thin, arid, and pale, and their top layer contains few plant nutrients. Darker, finer-textured, and richer in organic matter and plant nutrients are the characteristics of soils in lower layers. Base materials like calcium and magnesium are very poor in all laterite soils. The pH of the soil is low, and it has an acidic composition. Kaolinite with traces of illite is clay type of soil. The order of ultisols includes the majority of laterite soil types.

2.3 Desert Soil (Aridisols)

Hot desert is the region of desert soils in India. The majority of the rainfall, which can range from 50 mm to 400 mm, occurs during the monsoon season. Aridity is evident and potential evapotranspiration is very high. Sand dunes and undulations are a major part of the region. Common features of desert soils are a region where alkaline earth carbonate is present and where lime concretions are accumulating at a depth of 60–120 cm. The soil is more prone to dispersion and less permeable when sodium clay is present and clay content is ranging between 2–8% that is very low. The range of soil pH is 8–9. Illite clay with traces of kaolinite is the most common type.

2.4 Sub-Montane Soils

Sub-montane soils can be found in the undulating sub-Himalayan regions, which are covered in coniferous forests. Accumulation of organic matter, high rainfall, acidic nature, and absence of free lime are the features of regions with sub-montane soils.

2.5 Saline and Sodic Soils

Saline and sodic soils are present in arid and semi-arid areas of the nation where annual temperatures are high and evaporation rates typically outpace rainfall. In surface layers of the soil, the situation leads to accumulation of salts. The soils are salty and sodic with poor available nitrogen, zinc, and organic carbon and covering an area of 8.5 million hectares in India.

Sulfates and chlorides of calcium, magnesium, and sodium make up the majority of the soil type and saline soils contain excessively soluble salts. Total salt concentration is typically higher than 4 dS/m, as measured by soil EC. Aridity, saline groundwater, and water logging are all related to saline soils.

Carbonate and bicarbonate salts are commonly found in sodic soils. The exchangeable sodium percentage of sodic soils is more than 15% and soil EC is less than 4 dS/m. These soils are very hard when dry and very soft when wet, and they can be seen when muddy water stagnates after a prolonged period of rain. At high pH level (>9.5), these soils contain insoluble calcium carbonate.

2.6 Climate

Climate refers to the average daily weather conditions for a given location or region over a period of time.

2.7 Dry Climate

In a dry climate, the annual average precipitation falls short of evaporation. Arid and semi-arid climates are subcategories of dry climate.

2.8 Arid Climate

When the average annual precipitation is less than 500 mm, typically between 250 and 500 mm, a climate is considered to be arid. Rainfall is far too little to meet the atmosphere's evapotranspiration needs. To produce crops, the amount of precipitation is insufficient.

2.9 Semi-Arid Climate

More than 500 mm of precipitation falls on average each year in semi-arid climate and using dry farming techniques or additional irrigation crop production is possible in semi-arid climate.

2.10 Rainfall Characteristics

The production of crops is significantly influenced by rainfall in dry regions among the different climatic parameters as follows:

1. **Rainfall Variability:** Variability in rainfall is found both in time and space dimension. Coefficient of variation of rainfall is very high and there are significant seasonal variations in annual rainfall. In general, the coefficient of variation

decreases as rainfall increases. In regions with lesser rainfall, crop losses brought on by unpredictable rains happen more frequently. In India, the average annual rainfall is 1192 mm. India can be divided into four zones based on the average annual rainfall. India's total land area receives less than 750 mm of precipitation on average.

2. **Rainfall distribution and intensity:** In 3 to 5 rainy days, typically receives more than 50% of the total rainfall. Due to surface runoff, significant water is lost as a result of such heavy rain. Soil erosion is accelerated by this process. In the period when crops are growing, distribution of rainfall is more crucial in comparison to total amount of rainfall for dryland farming.
3. **Variations or anomalies in monsoon behavior**
   (a) **Monsoon's delayed arrival:** If the start of monsoon is delayed, it will be impossible to plant the crops or varieties that are advised for the area in time that results in uneconomical crop yields.
   (b) **Monsoon's early departure:** As compared to late onset of monsoon, it is equally or even more dangerous when the monsoon arrives early. Terminal stress occurs in rainy season crops that lead to poor crop yields. Because the soil moisture is not enough, post-rainy season crops fail particularly during the maturational and reproductive stages.
   (c) **Persistent dry periods:** 7–10 days monsoon breaks possibly not a serious issue, whereas, beyond 15 days breaks duration, particularly at crucial stages where soil moisture stress causes yield reduction. As compared to deep soils, crops grown in shallow soils may be more negatively impacted by drought caused by a break in the monsoon.

Constraints Limiting Crop Production in Dry Land Areas

3

Abstract

The productivity of agriculture in dry farming regions is low due to the cumulative effect of many constraints for crop production. The major constraints in crop production include: climatic constraints like rainfall, high atmospheric temperature, low relative humidity, hot dry winds, high atmospheric water demand; soil constraints like inadequate soil moisture availability, poor organic matter content, poor soil fertility, soil deterioration due to erosion (wind, water), soil crust problem, presence of hard layers and deep cracks; traditional cultivation practices; heavy problem of weed; lack of suitable varieties; and socioeconomic constraints like less access to inputs, non-availability of credit in time, the risk-bearing capacity of dryland farmer is very low, therefore, the dryland farmers resort to low input agriculture which results in poor yields.

Keywords

Constraints · Crop production · Productivity · Low input agriculture · Yields

3.1 Major Crop Production Limiting Constraints in Dry Land Areas

The majority of cropping is still done under rainfed conditions in the arid and semi-arid regions. Small farmers with limited resources make up a sizable portion of the farmers. Only low input subsistence farming with low and unstable crop yields is allowed due to the limited resource base. Due to the cumulative effect of numerous restrictions on production of crop, productivity of agriculture in dry farming regions low. The restrictions to production of crop can be classified into:

S. S. Walia et al., *Rainfed Agriculture and Watershed Management*,
https://doi.org/10.1007/978-981-99-8425-1_3

(a) Climatic restrictions.
(b) Constraints associated to soil.
(c) Conventional farming techniques.
(d) Significant weed issue.
(e) Absence of appropriate varieties.
(f) Economic and social problems.

Climatic Restrictions

Specifications of Rain

In dry regions, rainfall has a significant impact on crop production among the different climatic parameters.

Rainfall Variability

Variability in rainfall is found both in time and space dimension. Coefficient of variation of rainfall is very high, and there are significant seasonal variations in annual rainfall. In general, the coefficient of variation decreases as rainfall increases. In regions with lesser rainfall crop losses brought on by erratic rains are more common. In India, the average annual rainfall is 1192 mm. India can be grouped into four zones based on the typical rainfall every year. In India, less than 750 mm of rainfall occurs on over 30 percent of the country's land area.

Classification of India into Different Zones Based on Rainfall

Zone I (Very low rainfall area): In this zone, average annual rainfall is less than 350 mm. It occupies about 13% of geographical area.

Zone II (low rainfall area): In this zone, average annual rainfall is 350–750 mm. It occupies about 22% of geographical area.

Zone III (very low rainfall area): In this zone, average annual rainfall is 750–1125 mm. It occupies about 36% of geographical area.

Zone IV (high rainfall area): In this zone, average annual rainfall is more than 1125 mm. It occupies about 29% of geographical area.

Rainfall Distribution and Intensity

In 3 to 5 rainy days, typically receives more than 50% of the total rainfall. Due to surface runoff, significant water is lost as a result of such heavy rain. Soil erosion is accelerated by this process. In the period when crops are growing, distribution of rainfall is more crucial in comparison to total amount of rainfall for dryland farming.

Variations or Anomalies in Monsoon Behavior

(a) **Monsoon's delayed arrival:** If the start of monsoon is delayed, it will be impossible to plant the crops or varieties that are advised for the area in time that results in unprofitable crop yields.

(b) **Monsoon's early departure:** As compared to late onset of monsoon, it is equally or even more dangerous when the monsoon arrives early. Terminal stress occurs in rainy season crops that lead to poor crop yields. Because the soil moisture is not enough, post-rainy season crops fail particularly during the maturational and reproductive stages.

(c) **Persistent dry periods:** 7–10 days monsoon breaks possibly not a serious issue, whereas, beyond 15 days breaks duration, particularly at crucial stages where soil moisture stress causes yield reduction. As compared to deep soils crops grown in shallow soils may be more negatively impacted by drought caused by a break in the monsoon.

An Extremely Hot Atmosphere

The need for moisture in the atmosphere rises due to high atmospheric temperature which causes excessive evapotranspiration rates causing stress related to moisture.

Minimal Relative Humidity

Whenever moisture is limiting, high ET losses as a result of low relative humidity lead to moisture stress.

Hot Dry Winds

Desiccation of leaves due to hot dry winds results in moisture stress. During summer months, storms of dust and the loss of fertile soil are results of soil erosion brought on by high turbulent winds.

High Atmospheric Water Demand

Most of the year, potential evapotranspiration (PET) is greater than precipitation due to high atmospheric water demand.

Constraints Related to Soil

The different soil groups found in dryland areas include black soils, red soils, and alluvial soils. In different soil groups, the crop production constraints are different. The most common type of soil, alluvial, has less severe crop production issues than red and black soils. The various soil production restrictions include:

(a) **Insufficient availability of soil moisture:** The ability of soils in dry areas to retain moisture is poor due to shallow depth, particularly in alfisols (red soils), lack of rainfall, and low organic matter content.

(b) **Poor organic matter content:** Inadequate organic matter content: The majority of soils in dryland areas have very low organic matter contents (less than 1%) as a result of high temperatures and insufficient organic manure addition. Physical characteristics of soil that affect moisture retention are adversely affected by insufficient organic matter.

(c) **Poorly fertile soil:** Due to inadequate organic material deposition and the depletion of productive soil on top caused by soil erosion, arid soils have low fertility index. Nitrogen and zinc amount are less in most of the dry land soils.
(d) **Erosion (by wind or water)-induced soil degradation:** In India, among different land degradations to which nearly 175 mha of land is subjected, soil erosion is very predominant. The erosion leaves poor subsoil for crop cultivation by loss of top fertile soil.
(e) **Soil crust problem:** The plant population is affected by the development of surface soil layers with a hard texture that prevent seedling emergence in red soils. Due to high runoff, crumbling of the soil surface after rainfall decreases infiltration and storage of the rain.
(f) **Deep cracks and the existence of rigid layers:** Black soils' hard layers (called pans) and extensive cracking have an adverse effect on crop development.

Traditional Cultivation Practices

Farmers have evolved their current management techniques over time as a result of their extensive experience.

The practices for traditional management are:

- Driving a plow along a slope.
- Seeds being broadcast or sown behind a country plow, which results in a weak and irregular plant stand.
- Monsoon season sowing.
- Rainfall-based choice of crops.
- Minimal amount of FYM applications.
- Hand weeding.
- Mixed cropping.
- Harvesting by application of traditional methods.
- Traditional system of preservation.

Extensive Prevalence of Weed

In dryland areas, the invasion of weeds is the most significant problem. Weeds can grow well in conditions that are favorable for crop growth. The growth of crops is attempted to be stifled by the weed seeds, which germinate earlier than crop seeds. The weed infestation is severe in areas that are fed by the rain because of constant rain and a severe labor shortage. To reduce the decrease in crop yields, it is necessary to suppress weeds in the initial phases of plant development.

Absence of Appropriate Cultivars

The majority of cultivars of crops that can be grown in dry areas are designed for irrigation-based farming. There are no specific varieties that grow only in dry regions. Therefore, more work needs to be done to develop cultivars of crops specifically suited for dryland agriculture.

Socioeconomic Restrictions

Dryland farmers' economic situation is extremely bad because

(a) Fewer inputs available.
(b) Lack of timely access to credit.
(c) Dryland farmers have very little capacity to assume risk.

As a result, low input agriculture is used by dryland farmers, which produces low yields.

Drought, Its Different Types and Drought Management Strategies

4

Abstract

Drought is a temporary condition that occurs for a short period due to deficient precipitation for vegetation, river flow, water supply, and human consumption. Drought is of various types based on duration, nature of users, time of occurrence, and using some specific terms. Plants can adapt to drought conditions by plant modifications, i.e., structural or functional, to survive and reproduce in a particular environment. Plants survive and grow in water-deficient conditions mainly by drought escaping and drought resistance. The different drought management strategies are adjusting the plant population, mid-season corrections, mulching, weed control, water harvesting, and life-saving irrigation and use of wind breaks and shelterbelts.

Keywords

Drought · Deficient precipitation · Drought resistance · Escaping drought · Plant modifications · Drought management

4.1 Drought, Its Different Types and Drought Management Strategies

Drought is a transient state that lasts only a short while and is caused by inadequate precipitation for human consumption, river flow, water supply, vegetation, and other uses.

Drought was initially defined as an extended period without rain.

Ramdas (1960) defined a drought as an event where the seasonal rainfall deficit is more than twice the mean deviation.

S. S. Walia et al., *Rainfed Agriculture and Watershed Management*,
https://doi.org/10.1007/978-981-99-8425-1_4

According to the American Meteorological Society, a drought is a period of unusually dry weather that lasts long enough to cause an extreme hydrological imbalance in the area that is affected.

Droughts can be categorized based on their duration, user types, timing, and use of particular terms.

Based on Duration

(a) **Permanent drought:** Agriculture in the desert can only be done by irrigation throughout the entire crop season because the sparse vegetation is drought-adapted and is known as permanent drought.
(b) **Seasonal drought:** This occurs in climate zones with well-defined wet and dry seasons. Most arid and semi-arid areas fall into this category. Crop varieties should be grown for a period of time that coincides with the rainy season, and planting dates should be chosen accordingly.
(c) **Contingent drought:** This is a rare downpour. In general, areas with humid or sub-humid climates are more likely to experience it than other places. It usually only affects a small area, is erratic, and is short-lived.
(d) **Invisible drought:** This can also happen in areas where there is a lot of rain. When rainfall is insufficient to compensate for losses due to evapotranspiration, excessive water stress occurs within the soil, resulting in suboptimal yields. This usually occurs in areas with high humidity.

Based on User Relevance (National Commission on Agriculture 1976)

(a) **Meteorological drought:** When an area receives less than normal annual rainfall spanning a considerable amount of time (month, season, or year), condition is known as meteorological drought.
(b) **Atmospheric drought:** Low humidity and hot, dry winds are to blame for this, which is a common occurrence. Even when the soil is sufficiently moist, it can still happen. It alludes to a problem in which the amount of transpiration exceeds the amount of absorption in a short period of time during the hot hours of the day, causing the plant to wilt. When absorption keeps pace with transpiration, the plant wakes up again.
(c) **Hydrological drought:** Long-term meteorological drought leads to hydrological drought, which causes tanks, reservoirs, etc. to dry out as surface water is depleted. It leads to water shortages in all areas that use water. The basis for this is the relationship between water balance and overall irrigation to help plants to mature.
(d) **Agricultural drought (soil drought):** The discrepancy between crop evapotranspiration and the soil moisture available cause soil moisture stress that results in agricultural drought. Usually, it moves slowly and steadily. Thus, plants are capable of at least partial adaptation to increased soil moisture stress.

Precipitation levels are low or unevenly distributed in both space and time, which leads to this circumstance.

Based on Time of Occurrence

(a) **Early season drought:** This can be attributed to the late onset of monsoons or prolonged drought after early sowing.
(b) **Mid-season drought:** This is because there is a long period between two successive downpours and the prolonged dry period makes the stored moisture insufficient.
(c) **Late season drought:** It occurs when the rain stops too early and the mature plants are starved of water.

Additional Drought-Related Terms

(a) **Relative drought:** Drought conditions do not always exist in other crops when they do in one crop. Inconsistent soil moisture levels and plant selection are responsible for this. For instance, a situation might not be a drought situation for peanut farming even though it might be for rice crops.
(b) **Physiological drought:** Increasing soil concentration causes an increase in the osmotic pressure of the soil solution, such as soil with high salinity or alkaline soil, and refers to a state in which plants, despite the presence of water, are unable to absorb water from the soil. It is not due to lack of water supply.

4.2 Crop Plant Adaptation to Drought and Drought Management Techniques

Crop Adaptation

Drought adaptation is the crop's capacity to flourish when faced with water restrictions. In order to thrive and procreate in a specific environment, plants must modify their structural or functional characteristics through adaptation. By avoiding drought and developing drought resistance, plants can thrive and grow in water-scarce environments.

Escaping Drought

Avoiding dry periods is the easiest way for plants to adapt to drought conditions. Phytoplankton, a class of numerous desert plants, germinates at the start of the rainy season and is only active during the rainy season, where it has a lifespan of only 5–6 weeks. These plants are not drought because they lack mechanisms to overcome moisture stress. A safety mechanism is provided by germination inhibitors.

The most crucial adaptation for growing in arid climates for crops is a variety's capacity to mature before the soil dries out. But only a small number of crops have brief periods of growth to be considered short-lived crops. Certain pearl millet varieties ripen within 60 days after sowing. Short-lived legumes such as green gram, black gram, and cowpea are added to this group as well. Drought can occur within the 60-day harvest period, so in addition to being early, drought resistance is also required. A disadvantage of early cultivar breeding is that the yield decreases with the shortening of the period.

Drought Resistance

Through a variety of stress avoidance and stress tolerance mechanisms, plants can adapt to drought. These mechanisms provide resilience to drought.

4.3 Avoiding Stress

Stress avoidance is the practice of retaining water even in dry conditions while maintaining a favorable water balance and avoiding stress and its negative effects. Stress avoidance is primarily based on morphological and anatomical properties, which are the result of drought-triggered physiological processes (Levitt 1972).

Under drought conditions, it is possible to achieve a good water balance through one of the following methods: (i) Water conservation by limiting transpiration before or after stress occurs (so-called water savers) or (ii) by increasing the rate of water absorption sufficiently to replace lost water (water consumers).

4.4 Water-Saving Mechanisms

1. **Stomatal mechanism:** Between species, stomata behave very differently in terms of openness and closure. Numerous cereal species have short-lived morning stomatal openings that close for the remainder of the day. However, there are variations in this regard between crop varieties, as illustrated by the following instances. The drought-tolerant two varieties of oats opened its stomata more quickly in the morning when moisture stress is at its lowest and photosynthesis can proceed with the least water loss (Stocker 1960). An interesting observation made in the course of trials on the effects of soil moisture regime on different wheat varieties was that the stomata of a semidwarf variety remained open throughout the day, while those of the tall variety were open for only a few morning hours remaining closed thereafter even in the presence of suitable soil moisture and transpiration conditions (Shimshi and Ephrat 1975). However, water-saving mechanisms based on stomatal closure inevitably lead to decreased photosynthesis, which can cause "starvation injury due to drought" (Levitt 1972).
2. **Improved photosynthetic efficiency:** One way to overcome the photosynthetic limitation imposed by closing the stomata to increase the ability to withstand water loss through transpiration is to increase the rate of CO_2 assimilation at

specific stomatal openings (Hatchs and Slack 1970). Many important crops (sugarcane, maize, finger and foxtail millets, sorghum) (Hatch et al. 1967), Bermudagrass (*Cynodon dactylon*), Sudangrass, Bahia grass (*Paspalum notatum*), Rhodes grass (*Chloris gayana*) (Murata and Iyama 1963), and certain Atriplex species bind most of its CO_2 in C4 of malic and aspartic acids; and called C_4 dicarboxylic acid (C4) pathway. They have a primary carboxylating enzyme which has both a high potential activity and a highly effective for CO_2 with such an enzyme, high rates of fixation of CO_2 and with relatively restricted stomatal opening (Loomis et al. 1971). Species utilizing the C4 pathway are said to have a high CO_2 assimilation rate for a specific stomatal opening, higher temperature and optimal light conditions for photosynthesis (El-Sharkawy et al. 1968), and higher light saturated rates of apparent photosynthesis (El-Sharkawy and Hesketh 1965).

Another photosynthetic process which reduces water loss without a concomitant reduction in photosynthesis is found in a number of succulent plants, which keep their stomata closed during most of the day. These plants have the so-called Crassulacean acid metabolism (GAM) which enables them to fix large amounts of CO_2 as organic acids at night and convert it into carbohydrate during the day. As the primary assimilation of CO_2 occurs at night, the lower temperatures prevailing limit the amount of water transpired for a given stomatal opening. The pineapple, for example, which has the GAM type of photosynthesis, produces about the same amount of dry matter per year as sugarcane, but often uses only 10–20% as much water. GAM plants haven't been used for agricultural purposes aside from pineapples and specific Agave species grown for fiber and paper pulp (Loomis et al. 1971).

3. **Negligible cuticular transpiration rates:** The cactus is a prime illustration.
4. During moderate dryness, on the leaf surface, lipids are built up, reducing transpiration. This has been shown to occur in soybeans (Levitt 1972).
5. **Reduced Leaf canopy:** Reduced transpiration surface area is the main way xeromorphic plants can conserve water. Other than the conventional techniques for minimizing the aboveground components, possibly the most basic form of transpirational surface area reduction is leaf curling, or rolling when water is scarce, which is found in many grasses. It is a characteristic phenomenon that rolling leaves can reduce transpiration by up to 75% in xerophytes from the desert and by nearly 55% in semi-arid conditions (Stalfelt 1956).
6. **Leaf surface:** Different leaf morphological characteristics can affect plant survival during droughts by reducing transpiration rates. Common leaf characteristics include thick cuticles, waxy surfaces, and spines. In certain species, drought stimulates the production of epidermal hairs (Shields 1958). The mere presence of pubescence on the leaves is no longer considered as an advantage in reducing transpiration (Kramer 1959), unless it be thought increasing their albedo.
7. **Stomatal number and location:** Having fewer stomata helps to slow down water loss. In some species, stomata may be located in depressions or leaf holes, which, by obstructing air currents, can further reduce transpiration.

8. **The impact of awns:** Wheat awned varieties are prevalent in drier and hotter areas and produce better yields as compared to awnless varieties, particularly in conditions of drought (Grundbacher 1963). Awns contain chloroplasts, stomata, and photosynthesizing capabilities; they contribute 12% of the total dry mass of the kernels to the total plant dry weight (McDonough and Gauch 1959).

4.5 Accelerating Water Uptake

1. **Efficient root systems:** Plants that can withstand droughts have a variety of visible root systems. These root systems respond to a variety of soil conditions, including the length of time the soil is dry, the depth of the soil that is usually wet, the presence and nature of elements of soil, and the salinity of the soil. Plants adapt to dry conditions primarily through root development, rather than root remodeling (Shields 1958). The term extensive root system refers not only to root development in terms of depth and lateral development, but also to root density per unit soil volume and secondary hair root count (Oppenheimer 1960). There are considerable differences between cultivated plants in the extent, depth, and efficiency of their root system. The number of small branch roots per unit length of the main roots in sorghum is nearly two times higher as has maize, and it is assumed that this is one of the main reasons for its greater drought resistance. Another adaptation of some xerophytes is their speedy response to a rewetting of the soil by the rapid formation of secondary rootlets ("rain roots") which may appear only 2–3 h after a fall of rain (Evenari and Negbi 1962). An important characteristic of certain cultivated plants is their ability to respond quickly to fluctuations in moisture content by having roots that grow rapidly towards sources of available moisture. It has been shown experimentally that drought resistance of varieties of spring wheat is related to the rapidity of growth of the primary roots (Aamodt and Johnston 1936).
   Plants are also known to differ with respect to the ability of their root systems to penetrate relatively dry soils. At or below the wilting point, oats, wheat, and barley exhibit minimal soil penetration, while two range grasses—Sideoats grama (*Bouteloua curtipendula*) and Love-grass (*Eragrostis sp.*) penetrated the soil quite extensively (Salim et al. 1965).
2. **High root-to-top ratio (RIT):** An elevated root-to-top ratio is one of the best ways to adapt plants to drought (Killian and Lemee 1956). This is because the root growth rate greatly exceeds the shoot growth rate under such conditions. This reduces the transpiring surface area, while the individual plant's root system receives its water supply from a large area of soil. In fact, an increase in root: top ratio may actually lead to a higher total dry weight for plants raised in arid environments in comparison to equivalent plants raised in moist environments.
3. **Osmotic potential differences in the plants:** According to Levitt, 1958 the difference in soil moisture content (0.5%) that causes permanent wilting could provide enough water for a plant to survive for 6 days. In some cases, this difference could be what separates life from death.

4. **Turning water consumers to water conservers:** Water spenders tend to have very high transpiration rates due to their rapid absorption of water. However, once their rate of water absorption becomes low enough to sustain water loss, they tend to develop some characteristics of ware savers (Levitt 1972).

Drought Tolerance

Drought tolerance can be demonstrated either by reducing stress brought on by a lack of moisture or displaying a high level of stress tolerance when the plant is in fact exposed to a low water potential.

Mitigating Stress

Adapting to drought based on mitigating the effects of stress allows plants to continue having a large internal water potential throughout the drought. Consequently, it can preserve cell turgor and development while preventing direct or indirect metabolic harm caused by secondary stress from drought or dehydration (Levitt 1972).

1. **Resistance to dehydration:** The easiest way to avoid drought damage is to resist dehydration, preferably to the extent that turgor is maintained, and at least avoid cell collapse after turgor is lost (Levitt 1972).
2. **Prevention of leaf collapse:** In many instances, the principal effect of leaf adaptation to dry conditions, such as thick cell walls, is to prevent wilting or collapse of the leaf, rather than to ensure reduced water loss. This explains why xeromorphic types of leaves frequently transpire more, when amply supplied with water, than mesomorphic types (Shields 1958; Kramer 1959), and thus make possible an efficient use of water during the short periods of wet conditions which are typical for dry environments. As soon as the soil moisture-level again becomes critical, the stomata close and the morphological adaptations to reduced water supply mentioned above become operative.

Tolerating Stress

Resistance to metabolic strain

The greater the elastic dehydration strain [Up to a certain amount, cell dehydration is elastic and entirely reversible; after that point, however, it becomes plastic, irreversible, and hence harmful (Levitt 1972)], larger the risk of a resultant plastic metabolic strain. When dehydration occurs, there are mechanisms for tolerating an accompanying plastic metabolic strain.

Starvation

The greater the dehydration a plant can tolerate and still keep its stomata open the lower will be its dehydration compensation point. This kind of drought tolerance will be due to the guard cells' capacity to maintain their turgid state despite the wilting of the leaf cells, which is made possible by an increase in the guard cells' solutes. (Levitt 1972).

Loss of Protein

There are two systems that work together to prevent protein loss in drought-like situations: a faster rate of protein synthesis or a slower rate of protein dissociation. Compared to older leaves, young leaves are more drought-resistant and this is ascribed to their greater protein (Chen et al. 1968). In tests with varieties of winter wheat, for example, water stress caused a greater decreased in RNA content in a nonresistant variety than in the resistant variety (Stutte and Todd 1968).

Direct Drought-Induced Plastic Strain Avoidance or Tolerance

Some taller plants may exhibit minimal transpiration restriction during drought and can become air dry without suffering long-term damage. Grama grass, for example, may lose 98.3% of its free water without injury (Oppenheimer 1960). Exposure to a sublethal stress may cause an increase in resistance so that the plant is able to survive an otherwise lethal stress. This results in developing methods for maximizing hardiness of crop plants.

4.6 Strategies for Drought Management

Drought management strategies are divided into the following headings:

1. **Adjusting the plant population:** In dryland environments, plant populations must be lower than in irrigated environments. Under dryland circumstances, the rectangle planting design must always be used. The plant population can be adjusted if there is moisture stress brought on by protracted dry spells with insufficient moisture delivery.
   (a) **Increasing the space between rows:** Reduce the competition throughout the entire growing season by increasing the number of plants in the row and shortening the distance between rows. Therefore, it is better suited for conditions with insufficient availability of moisture.
   (b) **Expanding the intra-row distance:** In this situation, the plants are spaced farther apart so that they develop lavishly from the beginning. The crop will compete for moisture when it is in its reproductive stage. Therefore, when the moisture supply is restricted, it is less advantageous than above.
2. **Mid-season adjustments:** Mid-season adjustments are contingent management techniques used on the standing crop to improve the low soil moisture levels brought on by protracted dry spells.

(a) **Thinning:** The crop will be protected from loss by lessening competition if all alternate rows or every third row are eliminated.
(b) **Spraying:** Spraying water or 2% urea every week for up to 10 days will help to preserve the crops such as groundnuts, castor, and red gram, during extended dry spells.
(c) **Ratooning:** The ratooning can be carried out as a mid-season adjustment after the dry season is over in the case of crops like sorghum or bajra.

3. **Mulching:** Spreading a covering substance like mulch on the soil's surface will stop moisture from evaporating. Mulching conserves the crop during drought by maintaining soil moisture content.
4. **Weed control:** In dryland conditions, weeds compete with agricultural plants for various growth-supporting materials. The majority of weeds compete more fiercely with crop plants for soil moisture because they need more water. Weed control, especially during the initial stages of crop development, helps to eliminate the effects of dry spells by conserving soil moisture.
5. **Collecting water and providing life-saving irrigation:** It is the process of collecting runoff water during heavy rains and storing it in various structures. During extended dry spells the water that has been conserved can later be used for irrigation that will save lives.
6. **Utilizing wind barriers and safety belts:** Structures that block wind flow and lower wind speed are called wind barriers. Trees are planted in rows as shelterbelts to shield crops from the wind. The wind's windward side is the direction that it comes from. The direction of the wind is toward the leeward side. The shelterbelts are placed on the other side of the wind. They don't completely block the wind flow. Depending on the porosity of the shelterbelts, some of the wind goes through them while some of it deflects and passes over them.

   Without causing turbulence, this slows the wind down. Depending on how tall the middle tree row is, the shelterbelts' ability to defend an area. On the windward side, shelterbelts often protect against desiccating winds up to five to ten times the height, and up to 30 times the height on the leeward side. As the speed of the wind decreases, the amount of evaporation reduces too. This means that the amount of water available for plants increases. You can see the benefits of the shelterbelts distinctly in years of drought. Additionally, shelterbelts lessen wind erosion.

References

Aamodt S, Johnston WH (1936) Studies on drought resistance in spring wheat. Can J Res 14:122–152

Chen DS, Serid S, Katchalski E (1968) The role of water stress in the inactivation of m-RNA of germinating wheat embryo. Proc Natl Acad Sci U S A 61:1378–1383

El-Sharkawy MA, Hesketh JD (1965) Photosynthesis among species in relation to characteristics of leaf anatomy and CO_2 diffusion resistances. Crop Sci 5:517–521

El-Sharkawy MA, Loomis RS, Williams WA (1968) Photosynthetic and respiratory exchanges of carbon dioxide by leaves of the grain amaranth. J Appl Ecol 5:243–251
Evenari M, Negbi M (1962) The means of survival of some desert summer annuals. Plant water relationships in arid and semiarid conditions (Proceedings Madrid symposium UNESCO) pp. 249–259
Grundbacher FJ (1963) The physiological function of the cereal awn. Botanical Rev 29:366–381
Hatch MD, Slack CR, Johnson HS (1967) Further studies on a new pathway of photosynthetic CO_2 fixation in sugar cane and its occurrence in other plant species. Biochem J 102:417–422
Hatchs MD, Slack CR (1970) Photosynthetic CO_2-fixation pathways. Annu Rev Plant Physiol 21: 141–162
Killian C, Lemee G (1956) Les xerophytes: leur c'conomie d'eau. In: Ruhland W (ed) Handbuch der PJlanzenphysiologie. Springer Verlag, Berlin, pp 787–824
Kramer PJ (1959) The role of water in the physiology of plants. Adv Agron 11:51–70
Levitt J (1972) Responses of plants to environmental stresses. Academic Press, New York
Loomis RS, Williams WA, Hall AE (1971) Agricultural productivity. Annu Rev Plant Physiol 22: 431–468. https://doi.org/10.1146/annurev.pp.22.060171.002243
McDonough WT, Gauch HG (1959) The contribution of the awns to the development of the kernels of bearded wheat. Maryland Agric Exp Stn Bull 103:1–16
Murata Y, Iyama J (1963) Studies on the photosynthesis of forage crops: II. Influence of air-temperature upon the photosynthesis of some forage and grain crops. Jpn J Crop Scince 31(4):315–322
Oppenheimer HR (1960) Adaptation to drought: Xerophytism. In *Plant Water Relationships in Arid and Semiarid Conditions*. Reviews of Research UNESCO
Ramdas LA (1960) Crop and weather in India. ICAR, New Delhi, p 127
Salim MH, Todd GW, Schlehuber AM (1965) Root development of wheat, oats and barley under conditions of soil moisture stress. Agron J 57:603–607
Shields LM (1958) Morphology in relation to xeromorphysm. In: Lora MS, Gardner LJ (eds) Bioecology of the arid and semiarid lands of South-West. New Mexico Highlands University Bull, pp 15–22
Shimshi D, Ephrat J (1975) Stomatal behavior of wheat cultivars in relation to their transpiration, photosynthesis and yield. Agron J 67(3):326–331
Stalfelt MG (1956) Die stomatare trnspiration and die physiologie der spaltoffnungen. Handb Der Pflanzenphysiol 3:350–426
Stutte C, Todd GW (1968) Ribonucleotide compositional changes in wheat leaves caused by water stress. Crop Sci 8:319–321

5 Stress Physiology and Preparation of Appropriate Crop Plans for Dry Land Areas and Mid Contingent Crop Plan for Aberrant Weather Conditions

Abstract

Water stress influence the crop plants by affecting water relations, photosynthesis, respiration, anatomical changes, metabolic reaction, hormonal interactions, nutrition, growth and development, reproduction, grain growth and yield. To increase and maintain the crop yield in dry soil areas, the crop plan and practices that have been developed and suggested are: proper and timely tillage, early sowing, choice of suitable crop/good plant material, proper plant population, cropping system, fertilizer use, timely weed control, efficient rainwater storage and its use, alternate land use system and planning for aberrant weather hazard. In the situation of late onset of monsoon, transplantation must be done with arrival of rains in case of certain crops and alternate crops/varieties should be used. If there is dry spell immediately after sowing, it is advisable to replant the crop. In case of break in monsoon-mid or late season drought, ratooning and thinning, mulching, weeding, and intercropping practices should be done. In the situation of early withdrawal of monsoon, practices such as plant density reduction, use of surface mulch, saving irrigation, increase in frequency of intercultivation, and stripping of leaves can be done.

Keywords

Crop plan · Crop yield · Monsoon · Water stress

5.1 Influence of Water Stress on Crop Plant

(a) **Related to water:** Affecting the process of transpiration, mobility, and uptake will alter the condition of water. Due to an increase in atmospheric humidity, the delay in absorption behind transpiration leads to turgor loss.

S. S. Walia et al., *Rainfed Agriculture and Watershed Management*,
https://doi.org/10.1007/978-981-99-8425-1_5

(b) **Photosynthesis:** The decrease in photosynthesis is caused by moisture stress, chlorophyll levels, leaf surface area, and a rise in the amount of assimilates that saturate leaves.
(c) **Respiration:** Moderate drought causes an increase in respiration, but as drought worsens, water content and respiratory rate decrease.
(d) **Anatomical changes:** Less intercellular gaps and smaller cells; thicker cell walls; more mechanical tissue development. There are typically more stomata per leaf.
(e) **Metabolic reaction:** Lack of water influences almost all metabolic reactions.
(f) **Hormonal interactions:** Hormones that encourage growth, include cytokinin, indole acetic acid and gibberellic acid, decrease in activity, while growth-regulating hormones such as abscisic acid, ethylene, and others increase in activity.
(g) **Nutrition:** Nitrogen fixation, absorption, and assimilation are affected. The uptake of NPK is decreased due to a significant decrease in dry matter production.
(h) **Growth and development:** A reduction in fruit, stem, and leaf growth. If drought hits before flowering, maturation is delayed; if it hits after blossoming, maturation is accelerated.
(i) **Reproduction and grain growth:** Drought at flowering and grain development determine the weight of each grain and the number of fruits, respectively. The start of the panicle in cereal plants is crucial because dryness during anthesis may cause the pollen to dry out. When it comes to moisture stress, vegetative and grain filling stages are less vulnerable, but dryness during grain growth lowers output.
(j) **Yield:** The amount of dry matter has a significant impact on yield that is regarded as usable for harvesting. If the yield parts are both in the air and underground, the impact of drought is just as delicate as the overall rate of growth. The moisture stress during blooming is damaging if the yield is made up of seeds, as is the case with cereals. When the economic product makes up a minor portion of the dry matter and the yield is made up primarily of fiber or chemicals, mild growth stress will not negatively affect yields.

5.2 Preparation of Appropriate Crop Plans for Dry Land Areas and Mid Contingent Crop Plan for Aberrant Weather Conditions

Choice of Crops

In dry farming areas, main food crops include millets and pulses. Such dominance of food crops is natural in a principally subsistence farming system. The choice of crops for drylands is affected by the following factors:

- Amount and distribution of rainfall.
- Beginning of rainy season every year.
- Duration of rainy season.
- Soil characteristics including the amount of rainwater stored in the soil.
- Requirements of farmer.

The main objective of the AICRPDA's research has been to identify the most effective crop for each dry farming area. The crop selection criteria for dry farming areas include:

- Drought tolerance.
- Fast growth during the early growing season to tolerate harsh conditions.
- High yield genetic profile.
- To escape terminal drought crops of short or medium duration.
- Wide climatic variations adaptability.
- Fertilizer responsiveness.

Suitable Varieties Selection

Traditional local varieties still dominate in the majority of crops of dry farming areas. These local cultivars are preferred due to their exceptional drought tolerance. However, they tend to last longer and are more prone to shortage of moisture during ripening. Even under ideal rainfall conditions, they have a modest yield potential. Improved management strategies like nutrition supplements do not have much success with them. Criteria used today to select dryland crop varieties include drought adaptability, high production potential, nutrient supply responsiveness, high water usage efficiency, and moderate insect and disease resistance including endurance. All dryland crop-compatible varieties have been created in all dryland areas, demonstrating their high-yielding potential.

Choice of Cropping System

Choosing the right cultivation system should aim to maximize and sustainably use resources, particularly water and soil. The amount of rainfall, rainfall duration, and soil storage capacity affects cultivation systems. Here are some general guidelines for choosing a dryland cropping system. Potential cropping systems based on rainfall and soil characters are presented in Table 5.1.

The crop and cropping system chosen must be appropriate for the length of the growing season without causing moisture stress to the soil. Climate analysis is useful for determining the best agricultural strategies for various regions.

Table 5.1 Rainfall and soil characters based potential cropping systems

Soil type	Rainfall (mm)	Season of growth (weeks)	Storage capacity for profiles (mm)	Cropping system
As below	>900	>30	>200	Assured double cropping maize – Chickpea soybean – safflower
Entisols, deep vertisols, deep alfisols, inceptisols	750–900	30	200	Double cropping with monitoring maize – Safflower soybean – Chickpea groundnut – Horse gram
Alfisols, vertisols, entisols	600–750	20–30	150	Intercropping sorghum + pigeon pea cotton + black gram
Alfisols, shallow vertisols	350–600	20	100	Single rainy season cropping sorghum/maize/soybean
Deep aridisols, Entisols (alluvium)	350–600	20	100	Single cropping sorghum/maize/soybean in kharif/rabi
Deep vertisols	350–600	20	100	Single post-rainy season cropping sorghum

Adapted from (Reddy et al. 2016)

Monocropping

Only one crop is grown in arid farming regions, when the average annual rainfall is less than 500 mm. Light soils like Oxisols, Alphisols, and Inceptisols are used to grow the crop during the wet season. On the remaining moisture in the vertisols during the winter, the post-rainy season crop is grown. Groundnut on Alphisols and comparable soils in Andhra Pradesh's Rayalaseema region, sorghum or chickpea or wheat in black soils in Karnataka MP, Andhra Pradesh, and Madhya Pradesh are some of the most widely practiced monocropping systems.

Intercropping

Growing two or more different crops with specific row configurations during the same growing period is known as intercropping. It is being recommended and practiced in areas where the average yearly rainfall ranges from 600 mm to 850 mm. We recommend and practice intercropping. In such areas, at least one component crop will be successful in dry years and there is a good chance of getting two crops during good rainfall years.

Depending on the soil, the amount of rainfall, the market price, and the marketing infrastructure, intercropping is implemented in various regions of the country. Depending on the market price of the constituent crops, the intercropping system is modified. Due to the numerous advantages it offers, including those listed below, intercropping is a typical method in dry farming.

- Intercropping reduces the crop failure risk by providing an insurance policy against total crop failure caused by abnormal rainfall.
- It increases yields and income per acre per unit time compared to monocultures.
- It ensures stable production.
- It provides multiple products for both domestic and commercial use.
- In addition, when legumes are intercropped, fertility of soil is increased.
- Weed growth is prevented by intercrop canopy.
- As water, nutrients and light are efficiently utilized, resource use efficiency is enhanced.
- One crop helps the other one grow physically.

Disadvantages include:

Due to competition effect the yield may be reduced.
Allelopathy effect.
Diseases and pests infestation.
Large farms with significant resources might not be able to use it.

Cereal-Based Cropping Systems

Cereal-based intercrop systems supply the highest quantity of various nutrients to the soil. Tables 5.2, 5.3, 5.4, and 5.5 represent the cropping systems based on sorghum, pearl millet, finger millet, and maize, respectively, for drylands and their production potential in a normal season in India.

Table 5.2 Sorghum-based cropping systems for drylands and their production potential in a normal season in India

Cropping system	Rainfall	Region (India)	Sorghum grain yield of (*t*/ha)	Pigeon pea grain yield of (*t*/ha)
Intercrop system				
Sorghum + pigeon pea	More than 800 mm	Telangana, Marathwada, Vidarbha, Malwa plateau, parts of Gujarat	2–2.2	0.6–0.8
Sole crop			2.5	0.8
Double cropping				
Sorghum-chickpea	More than 800 mm	Black soils of A.P., Karnataka, M.P.	2–2.5	2 (Chickpea)

Adapted from (AICRPDA 2014)

Table 5.3 Pearl millet-based cropping systems for dry lands and their production potential in a normal season in India

Cropping system	Region	Grain yield of pearl millet (*t*/ha)	Grain yield of g. gram/chickpea/ pigeon pea (*t*/ha)
Intercrop system			
Pearl millet + pigeon pea	Karnataka, Maharashtra, Gujarat, Rajasthan, parts of TS, M.P., Haryana, Tamil Nadu	1.2–1.5	0.6–0.8
Pearl millet + green gram	Jodhpur	1.2–1.5	0.4–0.6
Pearl millet (sole)		2	
Pearl millet-chickpea	Varanasi	1.5–1.6	0.8–1

Adapted from (AICRPDA 2014)

Table 5.4 Finger millet-based cropping systems for dry lands and their production potential in a normal season in India

Cropping system	Region	Grain yield of finger millet (*t*/ ha)	Grain yield of ragi (*t*/ ha)
Intercrop system			
Finger millet + Soyabean	Bangalore	1	0.2
	Ranchi	2	0.4–0.6
Finger millet (sole)		2.5	

Adapted from (AICRPDA 2014)

Table 5.5 Maize-based cropping systems for dry lands and their production potential in a normal season in India

Cropping system	Region	Grain yield of maize (*t*/ha)	Yield of second/ companion crop (*t*/ ha)
Intercrop system			
Maize + pigeon pea	Bangalore, Akola, Udaipur, Varanasi, Jammu, and Hoshiarpur	2–2.5	0.4–0.5
Maize + Soyabean	Bangalore, Akola, Indore, and Dehra Dun	2–2.2	1–1.2
Maize + green gram		2–2.2	0.4–0.6

Adapted from (AICRPDA 2014)

Oilseed-Based Cropping Systems

In the intercrop systems, red gram and castor go very well with groundnuts as the principal crop which shown promise in the Anantapur, Solapur, Akola, Ranchi, and

Rewa regions, whereas groundnut + castor system is becoming popular in the Rajkot, Ranchi, and Bangalore regions. Oilseed-based cropping systems for dry lands are presented in Table 5.6.

Double Cropping in Dry Lands

In places with substantial precipitation (above 900 mm), a long rainy season, and a good soil moisture retention capacity, double cropping, whether by sequential or relay, is feasible. Additionally, rainwater gathered in agricultural ponds can be used for double cropping in order to plant a winter crop.

Double Cropping by Relay Cropping

Red gram (June–September) + Groundnut/ragi (June–January)
– Horse gram (September–January)

Red gram and groundnut are intercropped in a 6:1 ratio in June. Horse gram is interplanted in the spaces between rows of red gram in September, following the groundnut harvest.

Double Cropping by Sequential Cropping

- Ragi (finger millet)/sama (little millet)/pearl millet can be grown between May–September and horse gram can be grown between September and January.
- Sesamum/Groundnut can be grown between May and September and Horse gram can be grown between September and January.
- Green gram/Cowpea can be grown between June and September and Sorghum can be grown between October and January.
- Sorghum can be grown between July and October and Chickpea can be grown between October and February.

Table 5.6 Oilseed-based cropping systems for dry lands and their production potential in a normal season in India

Intercropping system	Region	Yield of groundnut (*t*/ha)	Yield of companion crop (*t*/ha)
Groundnut + red gram	Anantapur, Solapur, Akola, Ranchi, and Rewa	0.9–1	0.5–0.6
Groundnut + Castor	Rajkot, Ranchi, and Bangalore	2–2.5	0.6–0.8
Groundnut (sole)		2.5	

Adapted from (AICRPDA 2014)

5.3 Pulse-Based Cropping Systems

The pigeon pea + groundnut system supplies the highest calories, proteins, and fat per hectare. However, the highest calcium and iron are supplied by a pigeon pea + finger millet system. Pulse-based intercrop systems and double crop systems for dry lands, and their production potential are presented in Table 5.7.

5.4 Contingent Crop Planning for Unpredictable Weather Conditions

Aberrant weather is bad weather that deviates from the normal weather pattern. Rainfed agriculture frequently experiences this. Due to the occasionally unusual weather, crop output in drylands is unpredictable. With regard to time and place, rain is sporadic, irregular, and seasonally specific. The monsoon season in India is characterized by its whims and breaks that can last for a variety of lengths. The customary kharif crop, which makes up 25–30% of the total area under crops, is not sown as a result of the delayed monsoon. The biggest issue with agricultural output in the region is Rabi crop seeding, which is also prevented by the early monsoon withdrawal.

So far, the crop plan and practices that have been developed and suggested for the purpose of increasing and maintaining crop yield in dry soil areas have been summarized as follows:

Table 5.7 Pulse-based cropping systems for dry lands and their production potential in a normal season in India

Cropping system	Region	Yield of first crop (*t*/ha)	Yield of companion crop (*t*/ha)
Intercrop system			
Pigeon pea + groundnut	Anantapur and Akola	1.2–1.5	1–1.2
Pigeon pea + finger millet	Bangalore and Bhubaneswar	1–1.5	0.8–1
Pigeon pea (sole)		1.5	
Double crop system			
Green gram + safflower	Bijapur, Solapur, Indore, Udaipur, Ranchi, and Hisar	0.6–0.8	1–1.5
Cowpea + finger millet	Bangalore	0.6–0.8	2–2.4
Cowpea + maize	Bangalore	0.6–0.8	2.2–2.5
Black gram + Rabi sorghum	Bijapur, Solapur, Akola, and Indore	0.5–0.6	1.2–1.5

Adapted from (AICRPDA 2014)

1. Proper and timely tillage.
2. Advance sowing.
3. Choosing an appropriate crop /good plant material.
4. Proper plant population.
5. Cropping system.
6. Fertilizer use.
7. Timely weed control.
8. Efficient rainwater storage and its use.
9. Alternate land use system.
10. Planning for aberrant weather hazard.

Proper and Timely Tillage

Primary tillage is done in order to increase water intake and minimize weed infestation. Farmers do it by country plough. It has to be carried out as early as possible, prior to the sowing season. Keeping this in view, the concept of off-season tillage has been developed. The practice increases soil moisture intake and results in efficient weed control.

Shallow tillage in off-season with showers before the monsoon season helps to conserve moisture and reduces weed growth. In light soils, surface crust formation reduces seedling emergence and rainwater intake, thereby leading to more surface runoff. A light cultivation would be adequate to open such soils. Area with hard pans which tends to produce hard pans by persistent crop cultivation and areas with textural profiles, i.e., with more clay, as depth increase requires the deep tillage. In soils with textural profiles and hard pans, deep tillage aids in increasing water content. In situ moisture protection requires the soil to be opened up so that it can act as a barrier to rainwater flow. Tillage machines must be of suitable size and type corresponding to the power sources.

Advance Sowing

After the start of the monsoon, dry land crops must be sown as soon as possible. Under some situations, dry sowing is desirable. By ensuring greater moisture conservation, good seedling vigor, availability of a longer growth season, and prevention of pests and diseases, early planting contributes to better crop output.

Choice of Suitable Crops/Good Plant Material

Choosing appropriate crops and varieties that ripen during actual rainfall seasons will not only increase the yield of one crop but also increase the intensity of the cultivation. The ability to yield relatively good yields under moisture stress conditions is the most important criterion for choosing a variety for dry land. To put it another way, crop types for dry land locations should be short-lived, drought-tolerant, and high yielding, able to be harvested during periods of rainfall, and have enough moisture remaining in the soil profile support post-monsoon cropping. The depth of soil, available profile moisture, and amount and distribution of rainfall are some of the important factors playing an important role in choosing which crops to grow.

Systems of Cropping

The rainfed areas are generally monocropped having a cropping intensity of 100. Farmers can raise crop in monsoon season, or fallow fields in post-monsoon season, or conserve rainwater during the monsoon and take a crop in post-monsoon season on conserved moisture. However, the researchers suggested that the cropping intensity can be increased through intercropping or double cropping depending upon the weather situation taking the rainfall, soil type, and other meteorological parameters into consideration, "effective growing season" have been worked out for different seasons. Based on this information, the following cropping systems for the country have been proposed.

Double Cropping

Areas with a soil-mixture storage capacity of more than 20 cm of water for a double crop and more than 750 mm of annual rainfall are suitable for double cropping. Planning the early sowing and harvest of kharif helps in setting the Rabi crop in a timely manner. Due to thermal sensitivity, most of the Rabi crops can't be planted. Surface soil starts to dry after the harvesting of Kharif crop and the seeding of subsequent Rabi crop can sometimes be a challenge. Pre-sowing irrigation can be helpful in enhancing seedling emergence. The use of advanced seeding devices such as ridger and seeder can be beneficial in dry land areas.

Double Cropping System
Cluster bean—Bajra
Cluster bean—Gram

Intercropping

In the 1970s, as a result of high-yielding, short-lived, and inputs-adaptable crop cultivars, agriculture shifted to a production-oriented model. This led to the development of intercropping to increase production over time and space. With this in mind, research has been conducted to determine the most profitable intercropping system for various regions in the country. While doing so due importance is given to the most common crop of the region and to pulses and oil seeds.

Use of Fertilizers

When attempts are made to level and bund fields in dryland areas, it is the surface soil that is removed, leaving the country's dryland soils not only thirsty but also hungry because these soils are constantly eroding both horizontally and vertically. The outcome is that the fields are reduced in depth and completely devoid of plant nutrients, particularly N, P, K. That is why it is important to make sure you are adding all three major nutrients in the correct amounts. Fertilizer use serves a crucial function in drylands. Not only that its use insulates crops against moderate moisture stress as the root system would be more developed compared to unfertilized crops. Thus fertilizer use helps in more efficient use of moisture, the scarce commodity in dry lands. Because the soil moisture in drylands is limited, there is a scarcity of nutrients. Hence, the objective should always be to apply fertilizers drilled by bullocks or tractors. Many farmers in dryland areas believe that crop failure is more likely when fertilizer is used. However, according to recent studies, fertilizer aids in both the efficient use of soil moisture profile in dry land locations as well as the availability of nutrients to the crop. Inorganic and organic fertilizers should be mixed properly. Although the organic form contains little nutrients, it helps to improve the soil's ability to retain moisture. Not only does it provide yield advantage, but nutrients such as potassium also through changing the connection, influence the plant's capacity to withstand drought between the plant, soil, and water. This reduces the transpiration losses and increases the productivity per unit of water. For the maximum benefit from the applied fertilizer, there is need for proper placement and split applications of fertilizers for maximum fertilizer use efficiency, which can achieve.

Fertilizer recommendation for dry regions are as follows

Mustard: 40 N kgha^{-1} and 20 P_2O_5 kgha^{-1}
Chickpea: 15 N kgha^{-1} and 40 P_2O_5 kgha^{-1}
Green gram: 20 N kgha^{-1} and 40 P_2O_5 kgha^{-1}
Cluster bean: 0 N kgha^{-1} and 30 P_2O_5 kgha^{-1}
Bajra: 40 N kgha^{-1} and 20 P_2O_5 kgha^{-1}

Timely Weed Control

Off-season tillage, proper seed bed preparation, and timely sowing help the crop in combating weeds. Studies conducted in the project very clearly show that the first 25 days period in case of 100 days crops and 35 days period in case of 150 days crops are sensitive to weeds. Therefore, one or two weeding with blade harrow or country plough keeps the weeds well in control. The effects are still better when the practices of interculture operation with bullock drawn implements along with hand pulling of intra-raw weeds is adopted and thus increase crop yields appreciably. For this purpose, some implements like blade harrow, rotating weeders like dryland weeder are recommended. Though the herbicides are quite effective in controlling weeds but the high cost of these chemicals takes them beyond the reach of dryland farmers.

Rainwater Management

In the management of rainwater, two strategies are used:

(a) Agricultural rainwater management.
(b) The collection of rainwater into excavated farm ponds; and the recycling of rainwater on the donor site.

On Farm Rainwater Management

Rainwater enters the ground through the surface. Therefore, soil surface need be more receptive for rainwater. For this, we do inter-row and inter-plot water harvesting, sowing across the slopes and use of broad bed and furrows.

Inter-Row -Water Harvesting

Cultivation of dry land of crops under ridge furrow system not only helps to take a good crop stand but also helps to conserve soil effectively. In this system, rainwater is absorbed at the point of impact and is not discharged outside the field.

Inter-Plot Water Harvesting

Crops should only be grown on half of the land in dryland areas where rainfall (300 mm) is insufficient to meet the water needs of any dryland crop. The other half should be left uncultivated with plans in place to add runoff water to the cropped part. These principles guarantee that crops will produce fairly even in years of drought. This technique is very good for cultivation for arid-fruit trees like Ber, etc.

Harvesting Rainwater into Dugout Farm Ponds

Dryland areas do receive at least one heavy rainfall during monsoon season, which causes runoff particularly from the fields having undulating topography and crusting soil. Runoff in drylands varies from 1040 percent of rainfall. So effort has to be made to harvest at least part of this runoff in dugout ponds. In order to control the loss of water from these ponds, several sealing solutions have been tested and cheap and effective materials have been identified, for example, in light soils, cement: soil moisture of 1: 8 or plastering with thick coal tar spray on the surface area of the pond are fairly effective.

Recycling of Stored Water

Even though moisture stress can lower crop yields at any stage of growth, there are some stages, which are more critical than other. Water availability is meager in dry lands so it is essential to apply it at the right time to get the best return. Therefore, the critical stage of irrigation must be selected. For example, the flowering stage of sorghum in the rainy season is the most important and critical stage. In the case of tobacco, irrigation 3 weeks prior to topping gives the maximum pay-off. In the case of wheat and barley, the initiation stage of the crown root is the most sensitive stage. In comparison to short-lived shallow root crops, long-lived crops with deep root systems responded better to increased irrigation levels. To get best results, this indicates the importance of wetting the root profile.

Alternate Land Use System

Crop production is not possible in all drylands. In some drylands, range/pasture management is appropriate, but in others, tree farming, ley farming, and agroforestry systems are appropriate. All these alternatives for the production of crops are called as alternative land use systems. This technique reduces risk, utilizes off-season rainfall that would otherwise be squandered as runoff, eliminates soil degradation, and restores the ecosystem's equilibrium in addition to helping to generate the off-season employment that is required in monocrop drylands.

While drought-resistant grass and trees can be very profitable in drylands, crop productivity can be devastating during dry years. Alternative land use strategies that have been created by scientists studying drylands may be suitable for various agri-ecological circumstances. These systems, which make greater use of the available resources to boost and stabilize the productivity of drylands, include alley cropping systems, agri-horticultural systems, and silvi-pastoral systems.

Planning for Aberrant Weather Conditions

Rainfed agriculture frequently experiences erratic weather. The following four significant deviations from normal rainfall behavior have been more frequently noted:

(i) The beginning of the rainfall might be quite early or noticeably delayed.
(ii) Occurrence of dry spell immediately after sowing.
(iii) There may be extended gaps during the southwest monsoon season, when the majority of dryland crops are raised.
(iv) Rainfall may stop before the typical cutoff date or it may last past the typical rainy season.

Monsoon's Delayed Arrival

During certain times of the year, the southwest monsoon is delayed because of which the regular crops/varieties growing there can't be planted in time. The lower and even uneconomic crop yields result due to delayed sowing of crops. In such conditions, there are two management options available.

Transplantation

In the case of certain crops, community nursery must be raised at a location anywhere there is water, and when the rains start, transplantation must be completed. In delayed conditions, pearl millet transplantation has been proven to be beneficial, as the transplanted crop yields significantly higher grain yields than the direct seed crop.

Alternate Crops/Cultivars

When sown late in the season, certain crops have the capacity to produce superior yields even under relatively adverse moisture regimes, making them more efficient. Crops/varieties should therefore be selected based on the date of sowing. For instance, in the areas of west Rajasthan, short location crops such as green gram or cow pea performed well compared to pearl millet when sown late.

Dry Spell Instantly Following Planting

A dry spell may start right after the crop is sown in some years. As a result, the soil may crust, the seedlings may wither, and the crop stand may not establish well. To produce improved harvests, adequate plant stand maintenance is always required. Hence, in the event of a significant drought period happening right after planting, it is more advisable to replant the crop rather than persisting with an insufficient number of plants.

Break in Monsoon-Mid or Late Season Drought

(a) **Ratooning and Thinning**

The amount of solar radiation that the crop canopy blocks determines how much water the crop needs. The largest amount of leaf area development will take place 40–50 days after sowing, which will result in a rapid depletion of the soil moisture reservoir. Therefore, ratooning or crop thinning that results in a decrease in leaf area index somewhat lessens the negative consequences of drought. A drought-affected sorghum crop that was ratooned and then received a subsequent rain increased its grain output by 8 q/ha in comparison to 2 q/ha without ratooning. Ratooning is a high management technique and success that depends on the general vigor of the drought-affected crop.

(b) **Mulching**

If the interruption in the monsoon is of short duration, the application of soil mulching has been discovered as a technique to prolong the duration during which the earth conserves water. As a result, this helps mitigate the losses due to evaporation, thereby leading to the extension of the timeframe during which water remains accessible for use.

(c) **Weeding**

Weeds remove the soil moisture as well as nutrients; therefore, to protect the crop from drought, weed control is crucial.

(d) **Intercropping and Risk Distribution**

The extended interruptions in monsoon rainfall can lead to various levels of agricultural drought, impacting diverse crops to varying extents. Utilizing meteorological data about the frequency and likelihood of these interruptions, one can strategically choose a mix of crops with differing growth durations. This approach ensures that there is a time gap between their growth phases, enabling suitable intercropping arrangements.

Early planting of deeply rooted, drought-tolerant crops like pigeon pea followed by planting of other component crops may be helpful for regions with unpredictable precipitation in the early part of the season.

The companion crop should be shorter in length than the base crop, such as sorghum and green gram, for places with unpredictable precipitation in the latter half of the growing season.

In locations with more or less uniform rainfall, it would be excellent to produce an appropriate foundation crop together with a companion crop that would either have a longer or shorter lifespan depending on the period of growth.

For Hisar region, to minimize the risk the scientists suggested that 60% area for Kharif crops and remaining 40% should be allocated for Rabi crops and for successful Kharif system the 50% area for Bajra, 25% under Guar and remaining 25% area should be under short duration pulse/forage crops.

(e) "Fertilizer application aids crops in resisting drought by promoting the growth of a robust root system, enhancing the plant's ability to efficiently harness soil moisture."

Early Withdrawal of Monsoon

This type of situation is more dangerous in drought prone areas. When monsoon withdraws early then two types of difficulties are created. Rabi crops sowing may be jeopardized (are not sown in early Sept.). Moisture conserved in the soil is required to be carefully used by appropriate practices.

The suggested practices are as under:

(a) **Plant Density Reduction**
Rabi jowar should be planted in early September with a density of 1.0–1.35 lakh plants per hectare, requiring a 50% reduction. The plant population should be adapted before entering their vigorous growth phase (30–35 days after sowing).

(b) **Use of Surface Mulch**
By using organic surface mulches, moisture can be conserved in the soil. Five tonnes per ha mulch material is required for this purpose.

(c) **Saving Irrigation**
At the 55–56 day development stage, saving irrigation is typically administered. At 35 to 40 days of growth, the same may be used due to early withdrawal of monsoon.

(d) **Increase Frequency of Intercultivation**
Stopping rainfall prematurely leads to early soil cracking. To avert this cracking and the subsequent moisture loss, it is advisable to enhance the frequency of intercultivation. While the standard recommendation is three intercultivations,

opting for five or six can significantly aid in forming a beneficial dust/soil mulch.

(e) **Stripping of Leaves**
It helps to control moisture loss temporarily. This was not found to be useful practice for prolonged drought.

5.5 Adjusting Mid-Season: Crop Planning for Anomalous Weather Conditions in Dry Land Regions

According to the nature of rainfall, the following crops are grown in dryland areas:

1. Late monsoon arrival—setaria, sunflower, castor, red gram, horse gram.
2. Rains during July and sowing of kharif crops by the end of July or early August—basmati rice, mash, maize, maize fodder, baby corn, seasonal vegetables.
3. Rainfall in August and sowing through the month of August—maize, baby corn, maize fodder, bajra fodder, seasonal vegetables.
4. Late August rainfall and sowing up to early Sept—toria, maize fodder, bajra fodder.
5. Good monsoon onset—all kharif crops sowing.
6. Dry spell lasts more than 2 weeks—corrective actions.
   (a) A reduction in plant population.
   (b) A check on weed development.
   (c) An escalate in interculturing, in a drought-prone environment.
7. Early monsoon departure.
   (a) A reduction in plant population.
   (b) Surface mulch usage.
   (c) In a 30- to 45-day drought, protective watering.
   (d) Escalate the number of interculturing.
   (e) Cutting the leaves.
8. Monsoon extended—It does not happen very often.
   (a) Planting wheat rather than rabi sorghum.
   (b) In medium-deep soils, double cropping might be feasible.
   (c) Delaying the planting of rabi crops.

References

AICRPDA (2014) Annual reports of AICRP on dryland agriculture. Central Research Institute for Dryland Agriculture, Hyderabad

Reddy GS, Reddy YVR, Vittal KPR, Thyagaraj CR, Ramakrishna YS, Somani LL (2016) Dryland agriculture. ISBN: 978-81-8321-404-9

Tillage, Tilth, Frequency and Depth of Cultivation, Compaction in Soil Tillage, Concept of Conservation Tillage, Tillage in Relation to Weed Control and Moisture Conservation

6

Abstract

Tillage can be defined as the act of changing the soil's condition to ensure it becomes conducive for the growth of crops whereas soil tilth is a physical condition of soil, especially in relation to its suitability for planting or growing a crop. Tillage frequency depends upon texture of soil. However, tillage depth depends on crop, type of soil, and time of tillage. Moreover, by means of minimizing tillage activities, the aims encompass achieving the conservation of water, soil, and energy. Tillage in relation to weed control and moisture conservation comprises of minimum/optimum/reduced tillage, conservation/mulch tillage and zero tillage or no-till system.

Keywords

Tillage · Tilth · Tillage frequency · Tillage depth · Weed control · Moisture conservation

6.1 Tillage

The process of using tools and equipment for physical soil alteration in order to create optimal circumstances for seed germination, the successful establishment of seedlings, and crop growth is referred to tillage. Tillage can be defined as the act of changing the condition of soil to ensure that it becomes conducive for the growth of crops (Cuplin 1986). In arid regions, the goals of tillage include:

1. To allow rapid infiltration and good retention of rainfall create the soil's desired structure for a seed bed.
2. Soil erosion can be reduced by using tillage across the slope and contour tillage.
3. Eliminate undesirable agricultural plants and control weeds.

S. S. Walia et al., *Rainfed Agriculture and Watershed Management*,
https://doi.org/10.1007/978-981-99-8425-1_6

4. Manage agricultural remains.
5. To preserve in situ moisture, planting, drainage, etc., obtain specific land configurations.
6. Integrate and blend soil amendments, manures, fertilizers, or pesticides into the soil.
7. Achieve segregation through the relocation of soil from one stratum to another, eliminating rocks or extracting roots.

Consequently, it is essential to consider the tilling depth, tilling timing, tilling direction, and tilling intensity.

6.2 Tilth

Specifically in connection to the soil's suitability for planting or growing a crop, soil tilth is a physical property of the soil. The development and stability of aggregated soil particles, level of aeration, moisture content, soil biota, water infiltration rate, and drainage are all factors that affect tilth. Depending on environmental factors including variations in moisture, tillage, and soil additions, tilth can alter quickly.

Large pore holes in healthy soil allow for both air infiltration and water flow. Only in soils with sufficient soil oxygen levels may roots grow. A sufficient amount of water and nutrients are also present in this soil.

6.3 Tillage Frequency

It pertains to the frequency of tillage occurrences. In shallow soils with a light texture, regular ploughing will break them into a fine powder, heightening their vulnerability to erosion. For greater water retention, maintaining the land in a coarse and clumpy state before sowing proves beneficial for denser soils.

6.4 Cultivation Depth

Tillage depth—Tillage depth relies on crop, soil type, and tillage timing.

(a) **Deep tillage:** Implementing deep soil cultivation, at a depth of 25–30 cm, proves advantageous for enhancing the permeability of dense clay soils. This process aids in the sealing of fissures that emerge during the drying phase. In cases where the soil possesses compacted layers, performing heavy tillage about every 2–3 years, using a chisel plough set to penetrate 35–45 cm deep at intervals of 60–120 cm, can extend the viable rooting depth and augment moisture retention capacity. This approach is especially useful for crop cultivation such as cotton and red gram and crops which have deep-reaching roots. However, in shallow, light-textured, or gravelly soils, it is not advised to deploy.

(b) **Medium deep tillage:** For the majority of soils and crops, employing a tillage technique of moderate depth, around 15–20 cm, is generally satisfactory. This method is suggested for soils of intermediate depth, crops with shallow root systems, soils possessing a horizon without compacted layers, and for integrating crop residue into the soil.

(c) **Shallow tillage:** In soils with a light texture, shallow profiles and those particularly susceptible to erosion, shallow tillage, which reaches depths of up to 10 cm, is practiced. Shallow stirring or light harrowing proves beneficial in cases where soil surfaces are prone to forming a crust. Soil's ability to retain moisture from rain also rises with the increase in depth of tillage, transitioning approximately between 7–8% with shallow tillage and 9–10% with medium–deep tillage, and to 11–12% with deep tillage.

6.5 Compaction in Soil Tillage

Deep tillage softens the hard pan, promotes deeper roots, which facilitates in greater exploitation of retained soil moisture and applied nutrients from the profile. This allows rainwater to sink down into the lower soil layer where it is less likely to evaporate.

Deep tillage increases root depth, enhances infiltration and water storage, and eventually increases crop output in soils that are susceptible to compaction, undergo crusting, and have poor water infiltration capacity.

6.6 Concept of Conservation Tillage

By means of minimizing tillage activities, the aims encompass achieving the conservation of water, soil, and energy. Both approaches typically involve leaving plant remains on the surface of soil, and each procedure is carefully planned and carried out to guarantee that residue or live plants are continuously covering the soil. While the practices of conservation tillage can promote certain objectives associated with alternative farming, such as enriching soil organic matter and curbing soil erosion, it is worth noting that certain conservation tillage techniques might augment the necessity for pesticides. The alteration of soil properties caused by conservation tillage impacts growth of plant and diminishes fields' water runoff. Consequently, the soil covered with mulch remains cooler, and the area beneath the residue retains moisture. This has led to the successful implementation of numerous conservation tillage systems.

6.7 The Role of Tillage in Weed Prevention and Moisture Retention

In dry places, rain falls simultaneously across a vast area. To achieve timely sowing before soil dries up, the time between field preparations and sowing should be reduced. This requires rapidly completing extensive tilling across a large area. In this regard, relying on bullock strength and the old-fashioned wooden plough may not be beneficial. Mechanization of tillage operations and usage of tillage equipment that is more effective are warranted. For the conservation of in situ soil moisture, tillage also encompasses land shaping in dryland areas. Time and money can be saved by using tools that can do both shaping of land and tillage in one action. One procedure that combines land shape, preparation, and sowing can save a lot of time. Useful tractor-drawn equipment, such as a broad bed former cum seeder or a basin lister cum seeder which can simultaneously shape the land and sow can be employed.

Minimum/Optimum/Reduced Tillage

It refers to a tillage system that aims to perform the fewest possible tillage operations, i.e., those required for superior preparation of the seed bed, quick germination, and upkeep of the optimal plant stand. Along with saving money, energy, and time, it also aids in preserving moisture. These systems' objectives include the following:

1. Lowering the amount of labor and energy needed to produce crops.
2. Preserving soil moisture.
3. Creating the best possible seedbed instead of uniformizing the entire soil surface.
4. Minimizing compaction of field.

The benefits include:

- Decrease in soil compaction.
- Decrease in soil erosion.
- Increased water infiltration.
- Expanded fertility of soil as a result of crop waste decomposition.
- Lower costs of production due to fewer tillage operations.

Drawbacks include:

- Decreased germination of seeds.
- Reduced nodulation of root in some crops.
- In order to speed up the mineralization process, more nitrogen is needed.
- Need specialized tools and equipment.

Minimum Tillage Techniques

(i) **Row zone tillage:** Only the area around the crop rows is harrowed after the primary tillage with a plough.
(ii) **Plough plant tillage:** A specific planter is utilized for soil pulverization, seed sowing, and seed covering after primary tillage.
(iii) **Wheel track planting:** Tractors are utilized to pulverize, sow, and cover seed using their wheels, following primary tillage.

Conservation/Mulch Tillage

Through decreased tillage operations, the goals are to achieve soil, water, and energy conservation. Both methods typically leave agricultural by-products on the topsoil, and each procedure is designed to keep either growing plants or crop residue continuously covering the soil. Some of the objectives of alternative farming, including raising soil organic matter and preventing soil erosion, may be advanced by conservation tillage techniques; however, some of these techniques may result in a greater demand for pesticides. With conservation tillage, soil characteristics are altered in manners that influence growth of plants and lower field discharge of water. The effectiveness of many conservation tillage techniques can be attributed to the mulched soil's cooler temperature and moist soil surface.

Zero Tillage or No-till System

Primary tillage is entirely avoided, while secondary tillage is only allowed in the crop zone, making it an extreme form of minimum tillage. This strategy makes extensive use of machines and pesticides that have little to no residual effect on the crop that will be planted. The equipment should contain attachments for four operations like opening the soil for seed insertion, cleaning the narrow strip above the crop row, planting and covering the seed.

6.8 Merits

1. Soil biological activity is increased.
2. Owing to crop leftovers breaking down, soil organic matter content is increased,
3. Surface runoff reduced.

6.9 Demerits

1. Seed germination is poor.
2. For mineralization, higher dose of nitrogen is required.

3. Some voluntary plants and perennial weeds predominate.
4. Incidence of pest and disease is high.

Reference

Culpin C (1986) Farm machinery, 11th edn. Collins, London. [This is a comprehensive textbook on farm machinery]

7 Techniques and Practices of Soil Moisture Conservation (Use of Mulches, Kinds, Effectiveness, and Economics)

Abstract

Mulches are used to conserve the moisture in the soil. Mulch refers to any covering material used on the surface of soil to prevent evaporation. Various types of mulches include mulching by straw and stubble, soil, chemical, plastic, pebbles, live mulching, and vertical mulching. Among all the mulches, soil mulch is the cheapest. Mulches influence soil properties by regulating temperature, preserving soil structure, reducing soil salinity problems, increasing the water content of the soil, and reducing soil erosion. However, anti-transpirants are used to prevent moisture loss from plants. There are four types of anti-transpirants viz., stomatal closing type, film forming type, leaf reflectant type, and growth retardant type that prevent loss of moisture from plants by their different actions. A number of soil and crop management techniques are used to manage soil and crop. Soil management techniques comprise leveling and bunding, water diversion bund, weed control, mulching, and tillage. Moreover, rainwater can be conserved by storing more of rainfall in root zone, by checking loss of water through evapotranspiration and by water harvesting.

Keywords

Anti-transpirants · Evaporation · Management techniques · Mulch · Moisture · Soil properties · Rainfall

7.1 Mulch

Mulch refers to any covering substance applied on the soil's surface to stop evaporation. Mulch can be grown and used in the field, or it can be grown and changed before being used in the field. Mulch can also be grown or changed before being used or it can be processed, created, and transferred before being used.

S. S. Walia et al., *Rainfed Agriculture and Watershed Management*,
https://doi.org/10.1007/978-981-99-8425-1_7

7.2 Various Kinds of Mulches

(a) **Mulching by straw and stubble:** For conservation of moisture, mulches made of straw and other crop leftovers, such as peanut shells, cotton stalks, and so on, can be applied on the soil's surface. Mulches of straw lessen evaporation by reducing absorbed energy by the soil as well as its mobility above the soil. However, obtaining sufficient crop leftovers for use as mulches is a challenge.

(b) **Mulching by soil:** Using tools for tillage of surface soil like danthis, guntakas (blade harrows), etc., you can frequently agitate the soil to create mulch of soil, which is a thin surface of loose soil. By preventing the capillary action that would otherwise cause the soil moisture to rise, earth mulch applied to the top 5–8 cm of dry soil efficiently lowers losses by evaporation. By lessening the direct impact of the atmosphere, the mulch of soil also guards against deep fractures in soils (particularly black soils), which also results in less evaporation. In the absence of weeds, the numerous intercultivations carried out in rabi crops aid in lowering evaporation losses. The least expensive type of mulch is soil mulch.

(c) **Mulching by chemical:** Evaporation can be minimized about 40% by mixing hexadecanol-like chemicals. By creating a diffusional layer to evaporation, compared to untreated soil, the top layer of treated soil dries out more quickly.

(d) **Mulching by plastic:** For evaporation control mulching by plastic is very effective. The color of plastic mulches might be black or white. Black plastic mulch will absorb solar radiation and increase soil temperature, which will hasten the germination of winter crops like wheat, barley, etc. The evaporation of soil moisture will be slowed by white plastic mulches, which reflect incident radiation.

(e) **Live mulching:** This phrase refers to the surface layer that covers the soil in an intercropping system by the plant canopy. For example, sword bean + sorghum, forage cowpea + sorghum.

(f) **Vertical mulching:** This approach involves digging trenches that are 40 cm broad and 15 cm deep at intervals of 2–4 m across a slope, and then filling them with organic waste or stubble until they reach 10 cm above the surface of the soil. Infiltration is boosted in black soils, runoff is controlled, gathered in shallow trenches, and transferred to neighboring layers of soil.

(g) **Mulching by pebbles:** When the soil surface is covered with tiny stone-like pebbles. The cultivation of dryland fruit trees will benefit from this mulching. In addition to lowering evaporation, pebbles placed atop tree basins also make it easier for rainwater to infiltrate the basin.

As compared to the kharif season, mulching during rabi/summer months is more beneficial. Organic mulches promote crop growth by preserving soil moisture, especially in situations where soil moisture levels are declining.

7.3 Soil Properties Influenced by Mulches

1. **Temperature of soil:** Depending on the type of mulch used, mulches can have a wide range of effects on soil temperature. Black plastic mulches have the potential to raise soil temperature, while white or reflecting plastic mulches tend to lower it. Crop residues regulate temperature by lowering it in the summer and raising it in the winter. This results from the interaction of evaporative cooling and radiation interception. By lowering the temperature, the sugarcane garbage mulch will improve sugarcane sett germination during the summer.
2. **Structure of soil:** Surface mulches absorb raindrops, making them less likely to disperse and sealing soil pores, preventing crust formation. As a result, soil structure is preserved. Due to mulch breakdown, mulches also enhance soil structure.
3. **Salinity of soil:** Soluble salts move only up to a certain depth in dry land because of the lack of rainfall. As the water evaporates, the salts move back to the surface. The establishment of seedlings and germination might be affected by salt buildup in the topmost layers. Soil salinity problems will be reduced by mulches as they increase infiltration and reduce evaporation.
4. **Water content of soil:** By weeds' induced infiltration, decreased evaporation, and transpiration, the soil moisture content is improved. In addition, surface mulches impede the free passage of water vapor from the soil's surface to the atmosphere to raise the soil's water content.
5. **Soil erosion:** Soil particles size, water and wind velocity, and ease of soil particles movement by water and wind are all related. The majority of the time, wind does not erode particles larger than 0.84 mm in diameter, while water can erode them quite quickly. The mulches lessen the direct effect of raindrops when they fall on the ground. This prevents soil dispersion and the subsequent closure of soil pores, which results in less soil erosion.
   Among all the mulches, soil mulch is the cheapest.

7.4 Anti-Transpirants and Their Different Types

Anti-transpirants are any substances used to prevent water loss from transpiring plant surfaces. The term "anti-transpirants" is also used interchangeably with the term "transpiration suppressants." The most effective anti-transpirants can cut down transpiration losses by 30–40%. In total, there are four different kinds of anti-perspirants.

(a) **Stomatal closure style:** Stomata on the leaf surface are the primary conduits for transpiration. In small quantities, PMA (phenyl mercuric acetate) like fungicides and atrazine like herbicides act as anti-transpirants by blocking stomata. Photosynthesis in mesophyll is known to be inhibited by PMA. Despite reports of success from research conducted in glasshouses, their utility in actual outdoor settings is limited.

(b) **Film forming type:** Due to the creation of a physical barrier, waxy and plasticized components that create a thin coating on the leaf surface prevent water from escaping. Since these compounds also inhibit photosynthesis, their effectiveness is constrained. Anti-transpirants of the film-forming type should have the following desirable qualities: they must form a thin layer, be more impervious to the passage of water vapor than carbon dioxide, and the film must remain continuous and not rip. Thin film or thick film anti-transpirants can be used to create these films.
Hexadecanol, a thin film-forming substance.
Mobileaf, Polythene S-60, a thick film forming type

(c) **Leaf reflectant type:** These are the white compounds that coat the leaves, increasing their albedo (leaf reflectivity). They bring down leaf temperatures and the gradient of the vapor pressure between the leaf and the atmosphere by reflecting radiation, which lowers transpiration. The temperature of leaf is decreased by 34 °C and transpiration is decreased by 22–28% with 5% of kaolin spray. Additionally utilized as reflectant anti-perspirants are celite and hydrated lime.

(d) **Growth retardant type:** These substances cause the roots to grow faster and the shoots to grow slower, allowing the plants to decrease their surface for transpiration and withstand circumstances of drought. The root/shoot ratio is increased.
For example: Phosphon–D, Maleic Hydrazide (MH), Cycocel – (2-chloroethyl) Trimethyl ammonium chloride (CCC).
In general, photosynthesis is minimized by anti-transpirants. Therefore, their use is limited to preventing crop death when there is a severe moisture stress. The crop can make use of the subsequent rainfall if survives. Anti-transpirants can also help nursery plants recover more quickly from transplant shock. They are somewhat useful in horticulture crops and nurseries. Fruits are less likely to shrink after harvest when waxy materials are utilized.

7.5 Crop and Soil Management Strategies

Methods for Managing the Soil

1. **Leveling and bunding:** In semi-arid areas, open field cultivation is generally practiced. Since fields are not bunded in this system, there is considerable loss of water which could have otherwise been avoided, provided fields were bunded. In the semi-arid alluvial plains of Uttar Pradesh, where bunding alone, leveling alone, and bunding cum leveling, respectively, resulted in yield increases of 35,63, and 98%, it has been extensively shown how bunding and leveling of land can preserve soil moisture.
It has been proven to be useful to build graded bunds across slopes in soils with a slope greater than 0.8% but less than 6%. For different slopes, spacing of graded bund is presented in Table 7.1.

Table 7.1 Spacing between graded bunds for various slopes

Slope (%)	Vertical interval (m)	Slope (%)	Vertical interval (m)
5.0–6.0	2.2	2.0–3.0	1.7
4.0–5.0	2.0	1.5–2.0	1.5
3.0–4.0	1.8	0.8–1.5	1.3

Adapted from Gadkary (1966)

Soils with slope greater than 6% are highly susceptible to erosion. Such soils should be provided with bench terracing and used for growing trees and grasses.

2. **Water Diversion Bund:** To avoid serious soil erosion, it is essential to build bunds for water diversion. For soil conservation efforts, the first line of defense against outside water pouring into the field is a water diversion bund.
   (a) **Grassed waterways:**
   Waterways should be constructed to drain excess runoff to a safe and suitable site. A waterway's cross-section is computed using the runoff rate, the total catchment area, and the maximum permitted water velocity, which is determined by the kind of bed.
   (b) **Deep ploughing during summer:**
   Deep ploughing during summer particularly in soils with slowly permeable subsoil layer has been found to increase moisture storage in the root zone and consequently the yield. Kanwar (1968) reported that deep ploughing to 30 cm where subsoil had slowly permeable layer at Ambala and Ludhiana increased the moisture storage in soil and enhanced wheat yield by 14%.
   (c) **Stubble mulching:**
   In stubble mulching, crops are cut at a certain height above the ground surface and left in the field so as to maintain a cover on soil. Stubbles retard the velocity of flowing water on soil surface and thus permit greater time for intake of water in the soil. Thus, it increases moisture in the soil. Kanwar (1968) reported that incorporation of rice husk at the rate of 62.5 tonnes/hectare was discovered to be improved the moisture status and consequently yield of wheat significantly.
3. **Weed control:** One of the best ways to increase the amount of water accessible to crops has been found to be weed control. In addition to other aspects of crop productivity and the amount of water per unit of dry matter produced, weeds compete with crops for soil moisture and transpire more than the crops.
4. **Mulching:** By reducing runoff, increasing infiltration, reducing evaporation, controlling weeds, reducing soil temperature due to radiation shielding, heat conduction and trapping and evaporative cooling mulches have an impact on the water balance in the soil. For semi-arid areas soil moisture and temperature are of greatest importance.
   Where precipitation is limited and erratic and evaporation is high, good yields are obtained if precipitation is conserved effectively. Although plastic films generally resulted in higher water conservation, under widespread field conditions, they are challenging to control and relatively expensive.

Mulch ingredients include included leftovers from crops and other plant wastes like straw, stover, sawdust, and woodchips. These materials are inexpensive, frequently easily accessible, and promote invasion.

5. **Tillage:** Since inadequate residue production occurs in arid environments to result substantial moisture conservation of soil, "soil mulch" created as a consequence of repeated ploughing is more practical in dry areas. With the capillary pore or subsoil to surface soil continuity broken by the soil mulch or dust mulch, evaporation losses are reduced.

7.6 Crop Management Techniques

Intelligent integration of factors like soil, crop, crop rotation, rainfall, weather, and management techniques are key components of crop management. The crop management system should be functional and efficient, have a strong scientific foundation, and be both practical and affordable for the farming community.

Rainfall water can be conserved in the following ways:

1. **By storing more rainfall in root zone**
   (a) Tillage.
   (b) Contour farming of bunding.
   (c) Vertical mulching.
   (d) Subsurface barrier.
   (e) Addition of pond sediments of organic matter.
   (f) Addition of gypsum.
2. **By checking loss of water through evapotranspiration**
   (a) Mulching.
   (b) Weed control.
   (c) Use of anti-transpirants.
   (d) Crop thinning.
3. **By water harvesting**
   (a) Onsite water harvesting.
      (i) Collecting inter-row water.
      (ii) Water collecting between plots.
   (b) Water harvesting for recycling.

Increasing Water Storage in Root Zone

1. **Tillage:** Tillage opens the soil make it friable and thus it helps in reducing runoff and increases infiltration of rainwater in this way it conserves the maximum amount of rainwater. But it is to be taken into account that whether we should go for deep tillage or shallow. Deep tillage is better than shallow tillage because it conserves more moisture and helps in control of weeds.

2. **Contour farming bunding:** On slopy lands, when tillage operation like ploughing, seed bed preparation, seeding, interculture operations are performed across the slope as nearly on contours as possible. It develops numerous tiny water barriers and thus reduces runoff as well as soil loss.
3. **Vertical mulching:** Vertical mulching is a recent innovation which is found to be helpful in situ conservation of moisture for increased crop yields. This has been specially found to be suitable to black soils of Deccan plateau whose intake rate is very low. It consists of jowar stubbles persist in 10 cm-high ditches that are 40 cm deep and 15 cm wide. Such trenches spaced at 4–5 m increased crop yields 400–500% in years of drought and 40–50% in years of regular rainfall ever control.
4. **Subsurface barrier:** The practice is suitable for sandy soils where water is lost through percolation. The placement of subsurface barrier of asphalt (2 mm thick at 60 cm depth) resulted in increased up take of *N* by bajra plants. These were associated with higher WUE and yield of crop.
5. **Addition of pond sediments of organic matter.**
6. **Application of gypsum**: Infiltration rates are increased by application of gypsum in alkali—soils.

By Checking Evapotranspiration Losses of Water

1. **Mulching:** With respect to dryland, mulch is any material placed at soil surface with a view to conserve the soil water. There may be various types such as—straw or residue mulch, soil mulch, plastic mulch, and chemical mulch.
   (a) **Straw mulching can help in conserving soil moisture in the following ways –**
      (i) By checking loss of water from soil through process of evaporation because it reduces heating of soil from radiation and reducing wind speeds near soils surface.
      (ii) By reducing runoff and increasing infiltration of water.
      (iii) By checking weed growth.
   (b) **Soil mulching:** By preventing water from rising to the surface by capillary action in dry soil that is 5–8 cm heavy, surface mulch successfully lowered water loss compared to soil with an undamaged surface. Following kinds experiment and conclusion, a soil mulch for moisture conservation became standard procedure in dry regions and dicta such as two cultivations can replace one irrigation, became very popular.
   (c) **Chemical mulching:** Evaporation is reduced by 43% by mixing of a long chain alcohol, Hexadecanol, at ¼ inch of the soil, this material which is resistant to microbial activity remained effective for more than a year. As compared to the surface layer of untreated soil which creates a diffusional barrier to evaporation, treated soil dried out more rapidly.
2. **Weed control**: Since the beginning of time, it has been recognized that one of the most efficient ways to increase the amount of water accessible to the crops and,

consequently, to increase WUE is to suppress weeds. For every unit of dry matter generated, weed usually transpires more water than the plants they coexist with. According to estimates, removing weeds from a corn field uses as much water as supplying a complete irrigation system when it is most needed.

3. **Use of anti-transpirants:**
   (a) Any substance applied to a plant surface during transplantation with the goal of limiting water loss from the plant is an anti-transpirant. Anti-transpirants can lower transpiration in the following methods. a) By boosting the leaf's resistance to water vapor loss.
   (b) By reducing net leaf energy absorption and boosting leaf reflectance.
   (c) By using growth retardants to reduce top growth.
   Anti-transpirant may be of four types.
      (i) Stomata closure, e.g., PMA (phenyl) mercuric acetate).
      (ii) Film farming, e.g., plastic of waxy material.
      (iii) Light reflecting type, e.g., Kaolin.
      (iv) Growth retardant, e.g., CCC (2 chloroethyl trimethyl ammonium chloride).
4. **Crop thinning:**
   One of the key strategies for achieving optimum yield in low moisture circumstances is maximizing plant population. Crop thinning could be a mid-season adjustment for dryland farming. It was discovered that thinning the plant population by eliminating the third row from each plant was helpful in reducing the moisture stress.

Water Harvesting for Reuse

It is a method of runoff farming, which aims at increasing runoff and decreasing of rainwater infiltration rate. It is suitable in both sub-humid and arid environments, in addition to dry and semi-dry ones where annual rainfall is sufficiently high but along dung spells do occur during crop growth.

Water harvesting is of two types:

1. Onsite water harvesting.
   (a) Harvesting water between rows.
   (b) Harvesting water between plots.
2. Water collection for reuse.

References

Gadkary DA (1966) A manual on soil conservation. Department of Agriculture, Government of Maharashtra, Poona, India

Kanwar RC (1968) Some observations on the traps occurring near Ghanyara, H.P., India. Vasundhara 4:55–59

Seeding and Efficient Fertilizer Use

8

Abstract

Establishment of an optimum plant population depends on seed treatment, sowing at optimum soil moisture, sowing time, sowing depth, sowing method, and crop geometry. Furthermore, no agricultural system can endure if soil fertility is not preserved. Potassium status ranges from medium to high, phosphorus status is low, and nitrogen status is universally poor in dryland soils. According to the findings of a study on fertilizer use in dry farming areas, the fertilizer dosage should be minimal. There are various factors influencing fertilizer use efficiency, i.e., selection of cultivars or varieties; timely sowing; establishment of a sufficient crop stand; timing and technique of fertilizer delivery; practices for moisture conservation; and timely weed control.

Keywords

Dryland soils · Fertilizer use efficiency · Plant population · Soil fertility

8.1 Seeding

Establishment of Optimum Population

The creation of the ideal population relies on the following.

Seed Treatment

Seeds can be treated for a variety of reasons, including defense against insects and illnesses, the biofertilizers introduction, and the development of tolerance for drought.

S. S. Walia et al., *Rainfed Agriculture and Watershed Management*,
https://doi.org/10.1007/978-981-99-8425-1_8

Seed Hardening

It is technique to increase the seedlings' tolerance to drought. To make seeds more tolerant of drought, seeds are first soaked in a chemical solution and then dried. Germination and establishment are impacted by the immediate post-sowing soil moisture stress. The seedlings can survive seed hardening in moisture stress conditions. Before sowing, seeds undergo partial hydration during seed hardening, followed by dehydration. In chemical solutions with a set concentration, seeds are soaked for a predetermined amount of time. To get rid of extra moisture, seeds are soaked and then dried in the shade. Water is ingested by the seeds while they are soaking, but the germination process is not yet complete. Thus, the hardened seeds are prepared for germination. Immediately after being sown in damp soil, seeds begin to grow. The emergence of seedlings before the topsoil dries up is aided by such early germination.

Optimal Soil Moisture for Sowing

For sowing, it is necessary to receive an effective rainfall of 20 to 25 mm, which can saturate the soil up to 10–15 cm depth. The germination and establishment of seedlings are negatively impacted by moisture stress during or right after seeding. Sowing must be done as soon as feasible after receiving soaking rainfall in order to ensure enough soil moisture at that time. Tools and techniques for sowing are essential in this regard.

Sowing Time

The best time to sow is when there has been enough rain to completely moisten the seeds and when it has continued to rain after sowing. The week with a rainfall of at least 20 mm and a probability of a wet week following a wet week as well as a coefficient of variation under 100%, is the optimum sowing time in a rainfed location. In addition to ensuring the best harvest, timely seeding may also aid in avoiding pests. In Maharashtra, 30 lakh hectares of kharif sorghum are grown, with more than 70% of those being hybrid varieties vulnerable to shoot flies. The pest incidence can be reduced if the seed is sowed in early July.

Pre-Monsoon Dry Seeding

Sowing after rainfall is difficult in some areas where thick clay soils predominate because of the soil's extreme stickiness. Pre-monsoon seeding is done here in dry soil 2–3 weeks before the start of the monsoon season. Only when seeds receive the ideal amount of rainfall, they will stay in the soil and sprout.

The benefits of dry seeding before the monsoon include:

1. Advance sowing.
2. Consistent germination and successful establishment.
3. Instead of preparing the area for post-monsoon sowing, using the first rainfall itself to encourage germination.
4. Having a stress-free maturity before the monsoon season ends and early maturity.

The following factors determine whether pre-monsoon dry sowing is successful:

1. It is only advised for crops like cotton and sorghum, which are bold seeds.
2. The start date of the monsoon and the duration of rainfall after sowing must be used to determine advanced sowing based on rainfall data.
3. Hardening seeds is necessary for rapid germination and drought resistance.
4. The depth at which seeds are planted must be such that they won't sprout until they are sufficiently moist by rain.
5. To allow for seeding in dry land before the monsoon, off-season tillage is required.
6. It is important to prevent soil insects from damaging seeds.

Pre-monsoon sowing examples

1. With a sowing depth of 5 cm and seed hardening with 2% potassium dihydrogen phosphate or potassium chloride, pre-monsoon dry planting for sorghum on black soils is indicated 1–2 weeks before the onset of the monsoon.
2. Cotton dry seeding in black soils before the monsoon is advised at least 2–4 weeks before the start of the monsoon. The depth of seed sowing should be 5 cm, and potassium chloride, CCC (500 ppm), or DAP at 2% level should be used to harden the seed.

Optimal Sowing Depth

When the soil's top layer loses moisture, germination and seedling growth are affected, when seeds are placed at relatively shallow depths or on the surface. Assuring early and uniform germination and seedling establishment by planting at a depth where there is enough soil moisture available. The ideal sowing depth depends on the crop, particularly on the size of the seeds and the plumule's penetrating ability.

Coriander	7 cm
Cotton, maize	5 cm
Pulses, sorghum, sunflower	3–5 cm
Pearl millet and minor millets	2–3 cm
Sesamum	1–2 cm

Sowing Method

To achieve effective germination and seedling emergence in arid areas, it is crucial to put the seeds in a layer of damp soil. To prevent seeding in arid soil, it is therefore essential to sow as soon as rainfall occurs. It is crucial to plant the seeds correctly, neither too deeply nor too shallowly. The appropriate separation between the plants is also necessary for the development of an ideal population. The manner of sowing affects the geometry, density, and sowing depth. Using seed drills, behind-the-plow sowing and broadcasting are common sowing techniques used in arid regions. Seed dibbling and seedling planting methods are also used for crops like cotton, tobacco, and chilies. Each approach has benefits and drawbacks. Seed size, soil condition, available time, cropping system, geometry of crop, sowing depth, source of power, cost of sowing, etc. influence the seeding technique chosen.

Geometry of Crop

Land shape that each plant occupies is determined by the space between rows and plants. It is based on the crop's root structure, cropping system, and crop canopy size.

Crop geometry (cm) for different crops in dry areas as follows:

Maize- 45 × 30.
Bengal gram, sunflower, cotton (Arboreum)- 45 × 15.
Coriander, Sunflower varieties- 30 × 15.
Sunflower hybrids- 45 × 20.
Sesamum- 30 × 30.
Groundnut- 30 × 10.
Cotton- 45 × 30.
Cotton (intercropping)- (60 + 30) × 15 in paired row.
Red gram- 60 × 30.

8.2 Efficient Fertilizer Use

If soil fertility is not preserved, no agricultural system can endure. Potassium status ranges from medium to high, phosphorus status is low, and nitrogen status is universally poor in dryland soils.

8.3 Low Rates of Fertilizer Application

Farmers that practice dryland agriculture avoid applying fertilizer due to the unpredictable nature of rainfall and the ensuing uncertainty about crop performance. However, the outcomes of the trials showed that using fertilizer to produce dry land crops increased yield.

Poor yields may arise from lengthy pauses in the monsoon in kharif or rabi crops' limited root zone moisture being consumed before the plants reach reproductive stage if fertilizers are utilized to satisfy the requirements of dryland crops. On the other side, low fertilizer application rates will result in poor yields. This must be prevented by evaluating the production potential of various places and controlling the soil fertility levels by only adding the amount of fertilizer necessary to maximize the potential. So, irrigated agriculture uses more fertilizer than dryland agriculture.

8.4 Response to Nitrogen

Shallow soils will result in a less stable reaction to applied fertilizer because fertilizer usage and loss of fertilizer will be more sensitive in shallow soils. Erratic rainfall effects and low response to nitrogen were observed in shallow alfisols and vertisols, respectively. As rain started to fall, the vertisols reacted to the fertilizer more vigorously.

8.5 Response to Phosphorus

Nitrogen insufficiency is more severe than phosphorus deficiency. Although various crops have a reaction to phosphorus, it is not as obvious and widespread as the response to "N." Due to the high phosphorus fixation capability of dry land vertisols, the crop plants will have reduced access to phosphorus in the soil solution. This clarifies why vertisols do not respond to phosphorus because the amount of phosphorus applied rarely exceeds 25–35 kg phosphorus/hectare that is insufficient to match the soil's capacity to fix phosphorus and crop demand. As long as the meager crop yields are not improved by efficient strategies for conserving soil and moisture, and wholesome crop nutrition, the reaction of dryland crops to phosphorus will remain limited and marginal.

8.6 Response to Potassium

On a wide scale, unlike with irrigated crops, potassium has not been able to cause a noticeable reaction in dryland grains and millets. When the monsoon is normal, dry soils, in particular red soils and sandy loam soils, that are low in potassium may react to mild quantities of potassium.

8.7 Response to Micronutrients and Secondary

There have been noted iron micronutrient shortage and reactions to its application in crops like chickpeas and peanuts. Many crops are responsive to zinc and many locations suffer from zinc deficiency; for instance, peanuts, corn, pearl millet, etc. Oilseeds and pulses react to the calcium and sulfur among secondary nutrients.

8.8 Response to NPK (Balanced Fertilization)

The effectiveness of additional fertilizers is increased by balanced fertilizer application. If other nutrients are limited, there may not be much response to generally inadequate nitrogen. Phosphorus limits nitrogen response more frequently. The yield response is improved when nitrogen and phosphorus are used. Economical responses were only found with low levels of nitrogen in dryland regions with annual precipitation of <750 mm, where prolonged periods of drought are common during the growing season.

8.9 Alternate Sources of Nitrogen

Alternative sources of "N" include biofertilizers, green manuring, organic manures, the addition of legumes to cropping systems, etc.

8.10 Green Manuring

In drylands, this is typically regarded as being unprofitable. Farmers in the Khammam district of Andhra Pradesh grow green gram during the rainy months of June through July, gather the pods, plow back the remains, and gather the sorghum in September. It is practicable in vertisols with consistent precipitation. In areas with consistent rainfall, horse gram can be grown in October or November following sorghum or sunflower. As green leaf manure, Subabul, Gliricidia, etc., leaves can be employed.

8.11 Legumes Incorporation into Crop Systems

Legume should be incorporated into the cropping system because it is a biological nitrogen fixing crop. In a cropping system, legumes contribute 25 kg of nitrogen per hectare.

8.12 Organic Recycling

There is a ton of opportunity for composting organic wastes like millets and other crop leftovers, sugarcane waste, weeds, and other farm wastes to be used in agriculture. By increasing their ability to retain moisture, these organic wastes and farmyard manure together increase the productivity of soils, particularly sandy loam soils.

8.13 Rhizobium Culture

Rhizobium inoculation of legumes failed to increase "N" fixation in dryland settings most likely due to unfavorable environmental factors, particularly protracted periods of little soil moisture during the growing season, extreme heat stress, and a lack of nutrients.

8.14 Integrated Nutrient Management System

For sustained agricultural production, it envisions integrating the use of chemical fertilizers based on need with organic manures, crop wastes, biofertilizers, legumes in crop rotation, and green manuring.

Biofertilizers like Azospirillum are frequently discovered around the roots of grasses and cereals. Its usefulness for dry land settings is due to its high ability to fix nitrogen, minimal energy requirements, and widespread establishment in grain roots, endurance to high temperature (30–40 ° C), and prolific establishment in the roots of cereals. It can replace up to 20 kg of nitrogen per hectare and has been extensively studied in sorghum and bajra.

VAM: By enhancing phosphorus intake, Vesicular Arbuscular Mycorrhizae have been discovered to alter the yield of numerous crops, including finger millet, bajra, groundnut, sorghum, cowpea, and pigeon pea.

Phosphate Solubilizing Bacteria (PSB) attracted a lot of interest. Such as *Bacillus sp.*

8.15 Important Conclusions from Research on the Application of Fertilizer in Dry Land Agriculture

According to the research findings on fertilizer use in arid agricultural settings, the fertilizer level should be lowest. Local variations are less sensitive than improved cultivars.

The key conclusions of studies on fertilizer use in rainfed situations are as follows:

1. The amount of soil moisture influenced how fertilizers responded to crops. The reaction in shallow soils increases with increasing rainfall.
2. The reactions to nitrogen are consistent across the board, phosphorus responses are acquired on alfisols, and potassium responses are restricted.
3. The amount of moisture that has been held in the profile will influence how rabi crops, or crops which are those grown after the rainy season, react to fertilizers. The majority of the nutrients must be applied as a basal application in the soil at sowing.
4. Depending on rainfall, nitrogen can be applied in split applications to kharif crops. When top dressing is required but the soil moisture is insufficient, a second split may be prevented.
5. Some regions show signs of zinc deficiency.
6. Balanced fertilizer application improved yields in years with average rainfall.
7. Legumes respond well to fertilization with “P.”
8. Applying nutrients topically was not always beneficial.
9. Responses when rates are low about 25–30 kg nitrogen/hectare were the most cost-effective.
10. Crops respond to roughly 30 kg phosphorus/ha on heavy black soils.

11. About 25 kg of nitrogen per hectare is contributed by legumes in cropping systems.
12. An integrated nutrient management system seems to offer better financial rewards.

8.16 Fertilizer Use Efficiency (FUE) in Dryland Agriculture

It speaks of the kg of product per kilogram of fertilizer treated with plant nutrients.

Factors affecting the effectiveness of fertilizer use

1. Selection of varieties.
2. Prompt sowing.
3. Creation of an adequate crop stand.
4. Timing and technique of fertilizer delivery.
5. Practices for moisture conservation.
6. Effective weed control.
   (a) **Selection of varieties:** The duration of the growing season will influence the choice of cultivars. Traditional types have low output and a lengthy lifespan. High yielding and short-lived cultivars produce greater yields while using fertilizer more effectively.
   (b) **Timely seeding:** An essential agronomic aspect determining crop yield and, consequently, the reaction to fertilizer treatment is the timely seeding of a crop. Early sowing promotes healthy seedling vigor and a longer growth season, which results in a more effective use of fertilizers and larger yields. Because of a severe moisture stress in the soil in crops grown during the rainy season and those grown after it, delayed sowings in drylands significantly lower yields. In South India, the ideal moment to plant crops is between June and mid-July, and the appropriate time to plant crops after the rainy season is early September. According to Prasad and Mukerji (1988), when planting dates were moved from the middle of November to the first 2 weeks of January, wheat yields in all agroclimatic zones of the country decreased by more than 50%. The typical advice is to apply more fertilizer at a greater rate, which results in lower fertilizer use efficiency, with the objective to make up for the yield decline caused by the delay in sowing.
   (c) **Adequate crop stand:** Adequate crop stand: An insufficient plant population is one of the main reasons of poor yields in the majority of crops which is brought on by insufficient soil moisture availability and soil crusting. Only when the ideal plant population is offered for the effective application of the applied fertilizer dose a cultivar's genetic potential become realized. When applied to a population of 67,000 plants per hectare (winter maize), 200 kg nitrogen/ha produced the same yield as 100 kg nitrogen/ha applied to a plant population of 83,000 plants per hectare.
   (d) **Fertilizer application period:** All of the phosphate and potassium required for kharif and rabi crops is typically utilized as a baseline dose. On light soils,

split nitrogen applications have been found to be effective for kharif. Fertilizer meant for top dressing can be kept in case of rain is not obtained in a timely manner. However, because there are few possibilities for rain to fall after planting, all of the nitrogen must be added to the soil when the rabi crops are planted. The most efficient fertilizer rates are those that take into account the amount of moisture that has been retained.

(e) **Method of fertilizer application:** In terms of improving fertilizer use efficiency, placing phosphorus fertilizer close to the plant row outperformed broadcasting. It is more crucial for crops cultivated on soil that lose moisture in the post-rainy season. Typically, seed cum fertilizer drills are used to place the seeds.

(f) **Moisture conservation:** The most restricting element in dryland agriculture is soil moisture. Water use efficiency and fertilizer use efficiency work best together. For raising the fertilizer use efficiency, a number of moisture conservation techniques have been suggested, including vertical mulching, BBF, setline cultivation, and runoff recycling.

(g) **Effective weed control:** In dry fields, weed removal has a significant influence on the yield of crop. For moisture and nutrients, weeds compete with agricultural plants, lowering crop yields by 30–60%. Maintaining weed-free fields, particularly in the early stages of crop growth, has maximized benefits by making optimum use of the limited soil moisture, which has increased fertilizer use efficiency as well as yields.

Reference

Prasad LK, Mukerji SK (1988) Effect of sowing time and stage of cutting on the herbage yield and subsequent seed production of oat. Seed Res 16(2):218–221

9 Concept of Watershed Resource Management, Objectives, Principles, Problems, Approaches, and Components

Abstract

Soil, water, and plants are the three most crucial natural resources and a watershed is an important management unit for natural resources. Watershed management places particular emphasis on flood protection, sediment control, and optimizing crop productivity. The primary objective of the watershed is to enhance the socioeconomic conditions of the general populace residing in the basin. This is achieved by the augmentation of their income-generating potential and the provision of essential amenities, including electricity, potable water, irrigation water, and protection from the adverse impacts of floods and droughts. A number of constraints are also a part of watershed that includes physical problems, resource use problems, end problems, and socioeconomic and other problems. Watershed management program have different components, i.e., soil and water conservation measures, water harvesting, crop management, and alternate land use systems.

Keywords

Constraints · Natural resources · Socioeconomic conditions · Watershed management

9.1 Watershed Management

The three most crucial natural resources are soil, water, and plants. Due to their interdependence, these resources require a unit of management in order to be managed in the most efficient and practical way possible. A watershed is a crucial management unit for natural resources in this setting.

S. S. Walia et al., *Rainfed Agriculture and Watershed Management*,
https://doi.org/10.1007/978-981-99-8425-1_9

9.2 Concept of Watershed Management

A watershed is any geographic area where precipitation runoff is collected and discharged through a common point or exit. It is described as a measurement of space that encompasses all of the land's surface and directs runoff to a single point. It is equivalent to a catchment region or drainage basin. A watershed is the fundamental unit of development since it is a hydrologic unit that can be managed.

Watersheds with defined hydrological boundaries are thought to be suitable for organizing developmental programs since the state of water resources determines the outcome of the overall agricultural development process. It is crucial to have multiple watershed-based development programs in place alongside fundamental soil and water conservation practices. The actions for development must be carried out from ridgeline to outflow point. The goal of a watershed management program in drylands is to maximize the area's integrated use of water, land, and vegetation in order to sustainably raise the manufacture of food, fodder, fuel, and fiber, as well as to lessen drought, moderate floods, reduce soil erosion, and increase availability of water. Watershed management refers to the prudent management of water and soil resources within a specific geographic area to promote sustainable agriculture and reduce flooding.

The management of watersheds has been implemented as part of several programs of government of India. The watershed development concept was implemented in 1987 by the Desert Development Programme (DDP) and the Drought Prone Area Development Programme (DPAP). The National Wasteland Development Board (NWDB)'s Integrated Watershed Development Project (IWDP), which began in 1989, similarly sought to develop wastelands according to watersheds. The National Watershed Development Programme for Rainfed Areas (NWDPRA), run by the ministry of agriculture, is the fourth significant initiative based on the watershed idea. Watershed development projects are funded by the ministry of rural development through the DDP, DPAP, and IWDP.

The watersheds could be categorized based on size are as follows:

Micro watersheds: A few hectares to hundreds of hectares make up the micro watershed. These can be created inside the fields of crops.
Small watersheds: The drainage area of the watershed is a few thousand hectares.
Large watersheds: The river basins are considered as large watersheds.

9.3 Objectives of Watershed Management

The concept of watershed management is sometimes used interchangeably with soil and water conservation, while there is a distinction in that watershed management places particular emphasis on flood protection, sediment control, and optimizing crop productivity. The primary objective of the watershed is to enhance the socio-economic conditions of the general populace residing in the basin. This is achieved by the augmentation of their income-generating potential and the provision of essential amenities, including electricity, potable water, irrigation water, and protection from the adverse impacts of floods and droughts.

The objectives are:

- The consideration of watersheds as a fundamental unit for the sake of development and optimal land utilization in accordance with land capacities.
- One potential approach to mitigating the impact of floods is the implementation of small multifunctional reservoirs and other water storage structures strategically located near the headwater of streams and in areas prone to flooding.
- Sufficient provision of water resources to meet the demands of residential, agricultural, and industrial sectors.
- The mitigation of organic, inorganic, and soil pollution.
- The optimization of natural resource utilization to enhance agricultural practices and related livelihoods, hence ameliorating the socioeconomic status of the local populace.
- The proposal entails the augmentation of recreational amenities, specifically the development and enhancement of picnic and camping areas.

The objectives of watershed management program can also be described in symbolic form by the expression: POWER.

Here the letters symbolize the following:

P = Production of food-fodder-fuel-fruit-fiber-fish-milk combined on sustained basis.
Pollution control.
Prevention of floods.

O= Overexploitation of resources to be minimized by controlling excessive biotic interferences like over grazing.
Operational practicability of all on farm operations and follow-up program including easy approachability to different locations in watershed.

W = Water storage at convenient locations for different purposes.
Wild animal and indigenous plant life conservation at selected places.

E = Erosion control.
Ecosystem safety.
Economic stability.
Employment generation.

R = Recharge of groundwater.
Reduction of drought hazards.
Reduction of siltation in multipurpose reservoirs.
Recreation.

9.4 Principles of Watershed Management

- Maximizing the utilization of land based on its inherent capacity.
- The preservation of the nutrient-rich upper layer of soil.
- Mitigating the risks of sediment accumulation in storage tanks, reservoirs, and valuable agricultural areas.
- Safeguarding the presence of vegetation throughout all seasons.
- The on-site preservation of rainwater resources.
- The safe redirection of surface runoff to storage structures via grassed waterways.
- The implementation of measures such as stabilizing gullies and constructing check dams can effectively enhance the process of groundwater recharge.
- The enhancement of cropping intensity and land equivalent ratio can be achieved by implementing intercropping and sequence cropping techniques.
- Enhancing the sustainable utilization of marginal areas through alternative land management strategies. Utilize many options and combinations of agricultural, horticultural, forestry, and pasture systems.
- The implementation of water collection techniques for the purpose of supplementing and providing irrigation during off-season periods.
- The promotion of sustainable practices that maintain the ecological balance between humans, animals, and plants within the water system.
- Enhancing farm revenue by engaging in agricultural endeavors such as dairy, poultry, sheep, and goat farming.
- Enhancing infrastructural capabilities pertaining to the storage, transportation, and marketing of agricultural commodities.
- Ensuring income stability and mitigating risks during adverse weather conditions.
- Establishing small-scale agro-industries.
- Enhancing the socioeconomic status of farmers.

9.5 Watershed Problems

(a) **Physical problems:** By carefully examining the current maps, bad lands, steep slopes, vulnerable geological structures, etc. may be located. The weather and hydrological data can be used to pinpoint issues like excessive runoff, strong winds, and prolonged and intense rainfall.

(b) **Resource use problems:** It is important to identify issues including overgrazing, fire, shifting cultivation, unregulated mining, and poor road construction.

(c) **End problems:** Rapid identification is required of the last consequences of degradation of watershed, such as erosion of soil, high sedimentation, landslides, floods, droughts, and water pollution. It is possible to identify the frequency and severity of these issues by studying data such as history.

(d) **Socioeconomic and other problems:** Working on watersheds can be very challenging when facing financial issues. At the beginning of the stage, any significant issues should be recorded. Poverty, land tenure, a low level of innovation acceptability, a seasonal labor shortage, a lack of knowledge, etc., may all be examples.

9.6 Watershed Development Action Plan (Watershed Management Steps)

1. **Identification and selection of watershed**
   Using a field survey, the watershed's perimeter must be identified that moves up to the ridge line from the watercourse's lowest point. The size of the region might range from 100 ha to 10,000 ha.
2. **Watershed description**
   A minimum amount of data must be gathered on location
   Climate
   Slope, shape, and area
   Soil - hydrology, geology, biological, erosion level, physical and chemical properties
   Vegetation- cultivated and native species
   Land-based ability
   Current patterns of land usage
   Cropping system, cropping pattern, and management
   Adopted farming system
   Farming economics
   Resource of manpower
   Socioeconomic data
   Institutional and infrastructural facilities
3. **Evaluation of issues and selection of potential remedies.**
4. **Technology components designing.**
   (a) Methods to conserve soil and moisture.
   (b) Runoff collection, storage, and recycling.
   (c) The best cropping system and land use.
   (d) An alternative agricultural system and land use.
   (e) Other methods of land treatment.
   (f) Development of livestock and other related activities.
   (g) Groundwater augmentation and recharging.
5. Creating base maps of the watershed that include all geological, hydrologic, physiographic, soil, and projected development factors for each watershed component.
6. Cost-benefit analysis to show the predicted cost of each component activity, the project's overall cost, and the anticipated benefit.
7. Fixing the timeline to reflect the project's duration, the anticipated completion date for each component activity, and the departments or agencies that will participate in each component activity.
8. Observation and assessment to gauge the progress of project and offer changes, if necessary.

9. For site-specific problems, to find solutions, conduct research on farms.
10. **Organizational requirement:** The organization is a key element of the watershed development project. Problems with land use can only be solved in close cooperation with owners. Watersheds should be between 300 and 500 ha in size at the microlevel in order to facilitate this interaction, and a group of 10 such watersheds might be governed by one organizational entity only. The project's optimal implementation organization may be a watershed development agency at the unit level. The watershed development agency should include chosen locals because no project can be successful without the involvement of the community. The organizational prerequisites comprise.
    (a) Agency for the development of water sheds with diverse personnel.
    (b) Personnel training.
    (c) Farmers training.
    (d) Credit institution.
    (e) Village association / Farmers forum.
    (f) Government-free organization.

9.7 Watershed Management Program Components

The following are the primary parts of the watershed program:

Strategies to Conserve Water and Soil

These steps, in addition to collecting water, assist to increase the soil profile's availability of moisture and the ease of access to surface water for additional irrigation. The conservation measures on arable lands can be categorized into three groups based on the kind, nature, and cost of hydraulic barriers:

Hardware treatments.
Medium software treatments.
Software treatments.

Permanent Measures

These steps are taken to improve the physiography, relief, and drainage characteristics of the watershed with the goals of reducing peak flow rates, limiting soil erosion, and managing surface runoff. Rivers, terraces, and bunds are the permanent measures in the watershed management project.

Waterways: In order to dispose of runoff water safely, both with and without vegetation- grassed waterways.

Bunds: contour bunds—In permeable soils with slopes up to 6% that are ideal for regions with little rainfall (less than 600 mm).

Graded bunds—For poor permeable soils with slope of 2–6% and for soils having crust like chalka soils and suitable for high rainfall areas (more than 600 mm).

Terraces: Bench terracing: Bench terracing is appropriate for soils with slopes between 16 and 33%. Bench terraces lessen the slope's length and steepness. By installing bench terracing in Ootacamund, the erosion rate on land that slopes by 25% was reduced from 39 ton per hectare to less than 1.0 ton per hectare.

Semi-Permanent Measures

With vast field sizes, these are typically interbund treatments in traditionally bunded areas. They are utilized to reduce overland flow velocity. These actions might be lost for 2–5 years.

(i) **Small section / key line bunds:** At half the vertical bund spacing, on the slope, a short section bund might be constructed, but it must be repaired every 2–3 years.

(ii) **Strip Leveling:** Surface flow velocity can be decreased by leveling 4–5 meter wide strips of land above the bund across the main slope of land. Blade harrows can be used to level strips of land every 2–4 years.

(iii) **Live beds:** One or two live beds with a width of 2–3 m on a gradient or contour can be used. It is possible for the vegetation in the beds to be both perennial and annual.

(iv) **Vegetative or live barriers:** The runoff water can be filtered or the overland flow slowed down by a couple of densely planted grass or legume barriers along the bund and at the midpoint of the slope. As a plant barrier, khus grass is frequently advised.

Temporary Solutions

Each year, these in situ moisture conservation processes need to be revised or upgraded. Compartmental bunding, broad bed and furrows, contour farming, dead furrows, mulching, and tillage are examples of temporary techniques that have recently achieved widespread appeal.

Water Harvesting

Harvesting of water is the process of gathering water for later use and storing it. It is a technique for generating, gathering, preserving, and using local surface runoff for farming in dry and semiarid areas. The region of catchment, the facility of storage, and the command area are the three parts of water harvesting system. The portion of the land that receives rainwater is known as the catchment area. The runoff water is kept in the storage facility from the time it is collected until it is used. Water is used at the command area. With certain modifications, water collection is done in arid as well as semi-arid areas.

Inducing Runoff

Even in regions with annual rainfall averages as low as 50–80 mm, rainwater harvesting is still feasible. Rainwater was collected by ancient desert dwellers by diverting it into fields or cisterns as it was flowing down slopes. Small villages, homes, animals, and other small groups may benefit from the modest amount of runoff that has been accumulated across a broad region. Runoff is produced either by alterations to the land or chemical processing to enable the gathering of greater amounts of rainwater.

Water Harvesting Methods

Arid Regions

The crop should be able to mature in the catchment area with sufficient water, and the agricultural techniques must maximize the utilization of the water resources. Because they have deep roots and can use runoff water that is held deep in the soil and not lost to evaporation, perennial crops are frequently an excellent choice.

(a) **Water Spreading:** The little rainfall arrives as brief, violent storms in arid areas. Rapidly flowing water enters gullies before flowing to the sea. The area experiences water loss, and the quick runoff can frequently trigger severe floods in places that were undisturbed by the storm. Spreading water is an easy irrigation technique that can be used in this circumstance. Floodwaters are purposefully redirected from their natural paths and dispersed across nearby plains. Dikes, minor dams, brush fences, and ditches all serve to redirect or slow down the flow of water. Crops are grown on wet flood plains or in valley floods.

(b) **Microcatchments:** If the plant is completely encircled by a rainwater catchment basin, it can flourish in an area with insufficient rainfall for its survival. At the lowest point in each microcatchment, a basin about 40 cm deep is created., and a tree is then planted there. The microcatchment's runoff is stored in the basin.

(c) **Traditional Water Harvesting Systems:** The three most significant traditional water collecting technologies in Rajasthan are tanka, nadi, and khadin. An underground tank or cistern called a tanka is built to collect and store runoff water from a roof top, a natural catchment area, or a man-made catchment area.

Cement concrete or stone masonry is used to line the vertical walls, and 10 cm of concrete is used for the base. The Nadi or pond of village is built to store natural catchments' water and has capacity from 1000 to 6,00,000 L. Nadis have capacities between 1200 and 15,000 m^3. During the rainy season runoff water from rocky catchments is gathered in valley plains in Khadin, a distinctive form of land use system. After the water in a shallow pond has receded, crops are cultivated in the winter on the remaining moisture.

Semi-Arid Regions

There are several and even centuries-old methods of water collection used in semi-arid regions.

(a) **Dug wells:** Underground water has been gathered and stored in hand-dug wells, then elevated for irrigation. Due to dissolved salts, water often has low quality.

(b) **Tanks:** On the plains, tanks are used to collect the runoff water from the hill slopes and forests. The storage tank, tank bund, spill route, sluice, catchment area, and command area are the parts of a typical tank system. With the use of a bund, on the plains, a storage tank is used to collect and hold the catchment area's runoff water. To remove extra water and prevent tank bund breaches spillways are constructed at one or both ends of the tank bund. To allow water to flow into the command area under control, the tank bund's middle region is provided with a sluice.

Bricks, Kadapa slabs, stone pitches, red and black soils, cement-concrete, soil-cement, soil-cement, and polythene sheet are a few examples of the various types of lining materials.

Typically, clay soil linings are the most affordable. Rubber or plastic floats help to reduce evaporation losses in farm ponds, especially in arid areas. White plastic sheet is inexpensive and widely accessible. Technology for farm ponds is commercially viable.

(c) **Percolation tanks:** Water collects in large gullies or rivulets that are in motion. The area's water table rises as a result of pond water soaking into the soil. The wells that are located further down, percolation tanks are employed for additional irrigation.

(d)) **Farm ponds:** For collection and storage of runoff water, these are small storage structures. Depending on their design and appropriateness to various topographic situations, farm ponds are split up into different categories.

Excavated farm ponds are excellent for flat topography.

Embankment ponds are suitable for hilly areas.

Ponds along the embankment that have been excavated.

Excavated farm ponds are in three shapes: circular, rectangular, and square. Large amounts of water can be stored in circular ponds. To hold 30% of runoff, farm ponds between 100 and 300 m^3 in size can be constructed. High seepage loss is a problem for farm ponds with red soils.

Crop Management

Dryland research centers and state agricultural institutions have created a set of dryland crop practices that are location-specific for all crops and cropping systems that comprise:

(a) Choice of cropping systems and crops which considers the growing season's duration.

(b) Best time of sowing.
(c) Balanced fertilizer regimens and nutrient utilization for crops and cropping systems.
(d) Management of weeds.
(e) A set of procedures for unusual weather.
(f) Contingent cropping.

Alternate Land Use Systems

An alternative land use system is a land use pattern that is distinct from the current or traditional land use system. Alternative strategies for land use have the following benefits:

- By enhancing biological productivity and profitability, optimizing resource use.
- Conservation and enhancement of resource base quality.
- Combining livestock and crops.
- Improvement in overall quality of farm life.
- Reducing the reliance of agriculture on non-farm inputs.
- Creating the possibility of employment.

Agroforestry (silvipasture, agri-horticulture, agri-silviculture, and alley cropping), ley farming, and tree farming are commonly known alternate land use systems.

9.8 Agroforestry

Agroforestry can be defined as a comprehensive and self-sustaining approach to land management. It entails the intentional incorporation or preservation of woody elements alongside agricultural crops, such as pasture and livestock, either concurrently or consecutively within the same land area. This practice aims to address both the ecological and socioeconomic requirements of communities. Agroforestry can be described as a comprehensive term including several land use systems and technologies. In these systems, woody perennials are intentionally integrated into the same land management units as agricultural crops and/or animals, often in a specific order or temporal sequence.

Agroforestry systems exhibit a significant interplay between several components, encompassing both ecological and economic dimensions. The utilization of an agroforestry system is deemed more favorable compared to sole tree farming since the inclusion of intercropped annuals helps to regulate income during the initial stages of tree growth when they are not yet capable of yielding profitable produce.

The different agroforestry systems are:

Agri-Silviculture

This proposed land utilization strategy integrates the cultivation of perennial arboreal plants with the cultivation of annual arable crops. The practice involves the integration of crops and trees. The tree component provides many resources such as food, fuel, lumber, and green leaf manure. Class IV soils found in arid regions with an average annual precipitation of approximately 750 mm are considered to be well-suited for certain purposes.

One potential combination for intercropping is *Leucaena leucocephala* with Sorghum, while another option is *Sesbania aegyptiaca* with various types of pulses.

Silvi-Pastoral System

The silvi-pastoral system is a traditional land management practice that combines forestry and livestock grazing in a symbiotic manner.

The main purpose of this technique is to enhance the limited availability of feed. This system incorporates the integration of pasture and/or animals with trees.

The combination of Acacia, Cenchrus, and Stylosanthes species.

The combination of Sissoo, Cenchrus, and Stylosanthes is being considered.

Agri-Silvi-Pastoral System

The agri-silvi-pastoral system refers to a land management approach that integrates agriculture, forestry, and livestock production.

This system incorporates the integration of crops, pastures, and/or animals with trees. Deliberate introductions of woody perennials, particularly those with fodder value, are intentionally undertaken. In addition to serving as sources of fodder and fuel, these systems have the potential to contribute to food production and soil conservation.

Agri-Horticultural System

The agri-horticultural system refers to a method of cultivation that combines both agricultural and horticultural practices.

Fruit tree agroforestry represents a specific type of agroforestry system characterized by the inclusion of fruit-bearing trees. This agricultural practice is commonly referred to as the food-cum-fruit system, wherein short-duration arable crops are cultivated in the intervals between fruit trees. Several fruit trees that may be considered are guava, pomegranate, custard apple, sapota, and mango. Pulses are considered to be significant arable crops within this particular system. Nevertheless, it is worth noting that the cultivation of crops such as sorghum and pearl millet in the interstitial areas between fruit trees is contingent upon specific needs.

Horti-Pastoral System

The horti-pastoral system is a form of land management that combines horticulture and pastoralism.

The horti-pastoral system is an agroforestry approach that entails the integration of fruit trees into pastureland. The combination of guava, custard apple, and ber is suitable for integration into a horticultural pastoral system alongside grass species such as *Cenchrus ciliaris* (anjan), *Panicum antidotale* (blue panic), *Dicanthium annulatum* (marvel), and *Chloris gayana* (rhodes), as well as legume species including *Stylosanthes hamata* (stylo), *Stylosanthes scabra* (stylo), and *Macroptilium atropurpureum* (siratro).

Alley Cropping

In arable lands, food crops are cultivated inside the confines of alleyways created by rows of trees or shrubs. This agricultural practice is alternatively referred to as hedgerow intercropping or avenue cropping. Hedgerows are initially trimmed to a height of around 1 m upon planting, and afterwards maintained through regular pruning during the cropping period. This practice serves the purpose of preventing excessive shadowing and minimizing competition with food crops. The utilization of agroforestry systems is advised in humid tropical regions, primarily as a viable alternative to the practice of shifting farming. In the semi-arid parts of India, alley cropping is utilized as a means of providing feed during periods of drought. This is achieved by mulching the crop with hedgerow prunings, but it is important to note that this practice does not typically result in enhanced crop output.

The advantages of alley cropping include the following:

- The provision of green fodder during periods of low agricultural productivity.
- A higher overall biomass production per unit area compared to solely cultivating arable crops.
- The efficient utilization of off-season precipitation when no crops are present.
- The creation of additional employment opportunities during the off-season.
- The function of serving as a barrier to surface runoff, thereby promoting soil and water conservation.

Three distinct types of alley systems have been identified based on their objectives. These include:

- Forage-alley cropping
- Forage-cum-mulch system
- Forage-cum-pole system

Forage-Alley Cropping System

The forage-alley cropping system is characterized by the simultaneous consideration of crop output and fodder production. The tree species that are appropriate for the establishment of hedgerows are *Leucaena leucocephala*, Colliendra, and Sesbania. Pigeon pea and castor crops have been identified as viable options for cultivation within the alleys of Leucaena.

Forage-Cum-Mulch System

The forage-cum-mulch system involves the utilization of hedgerows for both forage production and mulching purposes. Loppings serve the purpose of mulching in the course of the agricultural season and are utilized as fodder during the off-season. Significant enhancements in the agricultural productivity of sorghum, groundnut, green gram, and black gram have been documented in various locations.

Forage-Cum-Pole System

The forage-cum-pole system involves the establishment of Leucaena alleyways at intervals of 5 m along the contours. Hedgerows are established by the process of direct seeding and are regularly pruned to a height of 1.0 m every 2 months during the crop season and every 4 months during the off season. In accordance with the established guidelines, it is permissible for a Leucaena plant to be cultivated into a pole at intervals of 2 m along hedgerows.

Tree Farming

The practice of tree farming refers to the cultivation and management of trees for various purposes, such as timber production, reforestation, and ecosystem restoration.

Trees have the potential to thrive and produce a substantial harvest in areas where cultivating crops for profit is not economically viable. Farmers in arid regions choose to engage in tree farming due to factors such as labor expenses, scarcity during critical farm activities, and recurrent crop losses caused by drought. Various multi-functional tree systems (MPTS) have undergone testing to assess their appropriateness and economic viability across many contexts.

The following tree species are suitable for regions with an annual rainfall of less than 500 mm: *Acacia nilotica*, *Acacia aneura*, *Acacia tortilis*, *Acacia albida*, *Prosopis cineraria*, *Prosopis juliflora*, *Pithecellobium dulce*, *Leucaena leucocephala*, and *Tamarindus indica*, among others.

Timber-Cum-Fiber System

The practice entails cultivating both trees and perennial fiber crops concurrently within a shared land area. The practice of intercropping Subabul with Agave has demonstrated higher profitability in the Bijapur region of Karnataka.

Ley Farming

This agricultural practice entails the rotational cultivation of legume forages in conjunction with cereals. The agricultural practice that incorporates a rotation system involving pasture (ley) for both grazing and conservation purposes is sometimes referred to as alternate husbandry or mixed farming. The system has a low level of danger in arid regions. The incorporation of *Stylosanthes hamata*, a legume fodder, into crop rotation has been found to enhance soil fertility and concurrently enhance sorghum productivity.

10 Factors Affecting Watershed Management and Impact of Watershed Management Program on Sustainable Agriculture

Abstract

Size, shape, slope, drainage, vegetation, geology, soil, climate, land use, etc., are considered as important factors that influence the watershed management. The watershed areas of the country are suffering from the following problems: (1) different degradation problems; (2) improper land use; (3) excessive cropping; (4) shifting cultivation; (5) slope cultivation; (6) overgrazing; and (7) human and animal pressure, etc., which cause severe soil erosion, desertification, siltation of harbor reservoirs and croplands, floods, droughts, fertility and soil moisture loss, and migration. The different management practices comprise hydrological, biological, bio-inputs, waste reuse, and land management. The impact and effectiveness of watershed initiatives such as Sukhomajri and Bunga have clearly shown that it is possible to embed ecologically sound rainfed agriculture in economically sustainable development. There are various ways to help in integrated watershed development.

Keywords

Management practices · Problems · Watershed initiatives · Watershed management

10.1 Factors Affecting Watershed Management

Each watershed shows distinct characteristics, which are so much variable that no two watersheds are identical. All these characteristics affect the disposal of water. Several characteristics, namely, size, shape, slope, drainage, vegetation, geology, soil, climate, land use, etc., are considered as important.

S. S. Walia et al., *Rainfed Agriculture and Watershed Management*,
https://doi.org/10.1007/978-981-99-8425-1_10

Size

The size of a watershed forms a basis for further classification into different categories as:

- Sub watershed (100–500 sq. km)
- Milli watershed (10–100 sq. km)
- Micro watershed (1–10 sq. km)
- Mini watershed (less than 1 sq. km)

The size helps in computing many parameters like precipitation received, retained, and drained off. Large the watershed, more the heterogeneity of the other characters.

Shape

Based on morphometric parameters like geology and structure, watersheds differ in their shapes. The typical shapes include round, triangular, and elongated. Shapes and ratios determine the length, which has an impact on runoff characteristics, including runoff time.

Physiography

The type of land, its height, and its physical characteristics all have a significant impact on the design of the operations involved in going green. For instance, plains in a populous area might only be used for agriculture, while a mountainous tract might be best used for trees.

Slopes

The land utilization, rain fall distribution and movement, and watershed behavior can be controlled by slope. The velocity of overland flow and runoff, infiltration rate, and thus soil transportation are affected by the degree of slope.

Climate

Climate is a determining factor for management, etc. It determines the flow characteristics and thus erosion behavior.

Vegetation

The choice of how to green the watershed and how it will be done are aided by detailed information on the vegetation. Knowing the native species provides a reliable reference for choosing plants and crops. It confirms what greenery can be grown and where. With care soil capabilities could be analyzed, compared, and profitably confirmed for management.

Geology and Soils

The nature of rocks and their structures determine size, shape, physiography, drainage, and groundwater conditions, whereas soil factors such as depth, nature, moisture content, and fertility determine crops. Together, soils and rocks have an impact on how water is stored, moved, and infiltrated.

Hydrology

To achieve the final goal of growing greenery in a watershed, the availability, quantity, and distribution of surface water are basic. Hydrology parameters aid in quantifying the amount of water that is available, being used, and being exploited in other ways. They choose where to put conservation structures and how they should be built.

Socio-economics

In the management of watersheds, statistics of people, their health, cattle and farming practices, and share of participation are important.

Hydrogeology

The need for groundwater is constantly growing. As a result, understanding groundwater resources is important for predicting their future availability in the context of cooperative water use.

Water Resources in India

With a geographical area of 328 million hectares, India receives 1194 mm of rain annually or 392 million acre meters of water. This might be rounded down to 400 m ha of rainfall, including any yet-unknown snowfall. About 75% of this 400 m ha of rainfall falls between June and September, the southwest monsoon season, while the remaining 25% falls during the following 8 months. Most of the water is absorbed by the soil, and some is lost.

Constraints of Watersheds

The watershed areas of the country are suffering from the following problems:

(1) Different degradation problems; (2) improper land use; (3) excessive cropping; (4) shifting cultivation; (5) slope cultivation; (6) overgrazing; and (7) human and animal pressure, etc., which cause severe soil erosion, desertification, siltation of harbor reservoirs and croplands, floods, droughts, fertility and soil moisture loss, and migration. It has been estimated that in India about 5334 mt of soil is eroded annually, out of which 29% is permanently lost to the sea, 10% is deposited in reservoirs, and 61% is transported from one area to another and loss varies from 5.3 to 8.4 m. It is predicted that one-third of arable land is likely to be lost in 20 years, if the present trend continues.

10.2 Land Degradation Problems in India

Area subject to water and wind erosion: 144.13 m ha.

Area degraded through special problem: 29.52 m ha.

Water logged area: 8.53 m ha.

Alkali soils: 3.88 m ha.

Saline soil including sodic areas: 5.50 m ha.

Ravines and gullies: 3.97 m ha.

Area subject to shifting cultivation: 4.91 m ha.

Ravines and torrents: 2.73 m ha.

Total problem area: 173.65 m ha.

Total flood-prone area: 40.00 m ha.

Total drought-prone area: 260.00 m ha.

10.3 The Different Management Practices are as Follows

Hydrological

Various analyses of different agroclimatic zones and subzones of the country have clearly shown that overwhelming majority of the production areas (including some in sub-humid and humid regions) are subject to water stress and drought conditions. This is due to not only low annual rainfall but also erratic distribution during the year and over the years apart from unfavorable rains and improper land use system. Realizing this, hydrological manipulations through cover management and structural (micro) measures had received priority attention.

No plant can survive/grow without a minimum quantity of water and of appropriate quality. Thus, conservation, augmentation, and regulation of the supply of moisture vis-à-vis various plant growth stages have been recognized as the most important aspect of the redevelopment program. Since the supply of rainfall is not under the control of human beings, only manipulation of the incident rainfall or groundwater could help achieve this objective to a great extent. Again all users, such as plant, animal, and man, can get water after it has routed through land surface and soil profile. Thus in situ moisture conservation techniques and in situ and near-site water harvesting system should be the prerequisites in watershed management plans.

In Situ Moisture Conservation and H_2O Harvesting System

Managing and making available moisture to the primary production systems as well as the plants which are used for the stabilization of the protection of degraded/degrading areas is prerequisite. For in situ conservation over extensive areas, mechanical barriers across the slope have in use on both arable and non-arable areas. These are effective to reduce velocity and erosive power of runoff and also promote absorption specially in immediate upstream and downstream areas. However, for uniform and good performance of plants in terms of regeneration and growth, uniform absorption and distribution of moisture over as large an area as possible are needed. This calls for micro-treatment, such as dead and contour furrows, racking, etc., in between the barriers across the slope. The practice of growing grasses and legumes in the intervening area goes a long way in distributing the benefits of conservation and augmentation of moisture over an extensive area.

Complete replacement of mechanical barriers by vegetative ones may be possible only at some sites where the slope is not very steep, overland runoff is not a very big problem, and the process of erosion is not very advanced. A more important and technically sound measure is to develop the network of micro water harvesting structures over extensive areas. This would directly enhance the possibility of quicker establishment of plants and their better growth and development. There is scope to improve the design specification and layout of such systems, especially by adopting some of the details from the traditional system, such as nullah plowing, nullah bunding, bunches, small ponds, and other models in use for generations in arid, semi-arid, and even sub-humid areas.

Cover Management

The moisture conservation and augmentation would not be lasting by themselves. All these measures need to be taken up with the complementary support of vegetative regeneration. In many areas, the pressure of livestock, small and marginal farmers, and landless laborers is great. These areas must yield number of produce such as fodder, fuel, wood, and small items for consumption by promoting appropriate agroforestry models with distinct thrust on improving hydrologic regimes. The relation between contributing areas to storage and submergence areas of micro structure needs to be critically studied and appropriate guidelines developed for varying site conditions. Similarly the interrelationship between volume and recharge for better analysis and the identification of indicators that will help develop design specifications. More knowledge of the root system and its role in expanding soil volume that improves watershed retention is necessary for the selection of plant species, and relevant norms should be supplied.

Biological

The initial thrust of biological measures has been to increase the vegetative cover in varied quantities. But in the mid-1970s stress on trees seems to have edged out other plant species, such as bush, shrub, grass, and other legumes. In degraded areas, establishing and promoting tree growth at a faster rate which can protect the land is difficult, at least for a period of 2 years. Hence the emphasis should equally be on grass and legumes so that a ground cover and quick incorporation of organic residues, including additional nitrogen, could be achieved. The emphasis should be on growing legumes and grasses, which not only enrich soil but also offer good fodder, which is in extremely short supply all over the country.

In respect of biological measures, mono-planting needs to be replaced by mixed plantation. For the selection of species their impact on the hydrologic regime should receive more scientific consideration. For degraded lands emphasis should be given to grasses and legumes with reduced number of trees per hectare. The strategy to achieve maximum dividend would include different innovative models of agroforestry. According to the need of the area, the model should include livestock management, duckery, viticulture, and sericulture to generate economic value, besides small-scale enterprises based on biomass.

Bio-inputs

Reasonable research evidences are available to open a new dimension in accelerating the pace of establishment and growth of vegetation in degraded areas, particularly arid lands, mixed areas, coasts, and those affected by industrial and urban wastes, and the application of bio-inputs such as mycorrhizae, rhizobium, and organic chemicals to overcome various limitations and increase tolerance of the plants to water stress, chemical imbalances, etc. In the watershed management program, the application of these new bio-inputs has yet be initiated and popularized.

Waste Reuse

The sewage and garbage from the urban settlements and industrial affluent are being disposed through dumping on the low-lying areas or wet lands and discharging into water bodies as well as agricultural and other lands. These have been not only polluting water and degrading the environment but also rendering large areas unproductive. On the other hand, the nutrients and fresh H_2O available in these waste products have been successfully used by many in greening and also growing many consumable items. As there are contamination hazards, watershed management program may at least initiate utilizing waste H_2O and garbage as sources of nutrients and H_2O for redeveloping degraded areas where such risks don't exist or can be utilized.

Land Management

Land management is another very crucial factor in watershed management. Proper land management decreases the runoff loss, increases H_2O recharge, and keeps the soil fertile. The management practices are:

(i) **Plowing:** Plowing uphill increases runoff, which reduces soil moisture and eventually erodes the soil. With the plowing along the contours, soil erosion can be reduced up to 50%.

(ii) **Furrowing:** Another method of measuring consumption is to dig down for two to ten lines. This method produces good results when used with intercropping and hedging. Minor furrows are left by plowing, and rainwater flows are collected. For a selected spacing according to the scope, double plowing along the row speeds up the procedure even more. If planting and hedging are done, this method contains more than 50 mm rainfall.
ICRISAT has successfully experimented with a broad bed and furrow system (BBF), which consists of graded, wide beds divided by a furrow draining into grassed water channels. For best performance, the BBF system is set up on grades 0.4–0.8%. The soil is stabilized by the high location, which also serves as a bund, while change and erosion are caused by the shallow furrows.

(iii) **Trenching**
Another method of conservation is confined excavation following the contours. In areas with low slopes or abundant rainfall, the optimal dimensions are 0.3 m width and 0.6 m depth. In generally sloping fields or in regions with frequent light showers, the proportions can be switched. Higher trenching dimensions should be used where there is a significant degree of erosion caused by soft soil, heavy rainfall, or high runoff and velocity. When plants are cultivated along the trench to take advantage of the high moisture content there, trenching proves to be an effective conservation method

(iv) **Bunding**
Erosion can be controlled by bunding along the contours. Bunding is a common practice in regions with minimal rainfall and well-drained red soils. The system of streams that already exists is used as waterways. In certain steep sloping lands, the only choice is furrow bunding. Compartment, inter, and graded are the three kinds of bunding.

Bunding with a 0.2% gradient is referred to as graded bunding and is used to monitor overland flow. Utilizing the already-existing streams, a waterway system is created to accommodate storm- and rainwater. Gully stabilizing methods are also used to conserve soil effectively. By reversing the direction of water movement between the beds, gaps are intentionally left to conserve more rainwater in interbunding. The computation in this technique needs to be exact. For 100% rainfall harvesting in semi-arid and desert regimes, compartmental bunding involves bunding the areas perpendicular to contour. In areas with little rainfall, it is a great method for cultivating vegetation.

(v) **Gradoni**

Gradoni are slender ditches constructed along contours for capturing overland flow and boosting soil moisture. They are designed for certain plants and might be continuous or intermittent. On steep slopes of 30–70% reverse sloped trenches whereas in slopes up to 30% level trenches are built, and short or individual gradoni are constructed for dissected slopes with undulating terrain. Continuous gradoni, also known as minor trenching, are created for uniform slopes.

(vi) **Hedging**

In addition to addressing the issue of conservation, the approach also generates biomass and further stabilizes the ground through its root system. There aren't any additional expenses. Subabul hedging generates good biomass for providing nutrient-rich feed for the cattle despite the fact that vetiver grass is highly advised. Catching rainwater and increasing soil moisture are both accomplished through hedging. Additionally, it creates a wind buffer that, on a local level, can confine even storm winds. In order to boost crop yields, hedging is done on gently to moderately sloping farms. Hedging is cultivated using trenches, bunds, or furrows. To reduce the impact of flooding, they are cultivated on both sides of the stream. They virtually always save all of the rainwater in places that are prone to drought when used in conjunction with other techniques.

(vii) **Terracing**

In Jamaica's highlands, bunding and terracing are widespread practices. The management of the green cover has an impact on terrace effectiveness. Bench terraces are a group of regular, continuous level strips that are supported by steep banks and spaced vertically across the slope. Earth is used to construct the banks, which are covered in grass or easily accessible racks. It is useful for irrigation that uses rainwater and allows for interflow between downpours. Outward-sloped benches allow water to flow freely without benching the terraces in areas with heavy rainfall.

Sloped terraces have the benefit of being less expensive for banks and raisers. Another arrangement of level benches and outwardly sloping terraces is known as intermittent terracing. Level benching and onward-sloping berth terraces are combined in hillside ditches. For crops that are semi-permanent rainfed crops, this method works well. Depending on the soil profile and water requirements, fruit plants can be planted in basins or pits. In this strategy, crops that need good transportation routes are grown in places with heavy rainfall using a hexagonal design. Bench terraces surround the basin in convertible terraces. Mixed farming is a perfect fit for this method. All of these terraces have vegetation or crops protecting the slopes.

(viii) **Measures Against Shifting Cultivation**

The agro-horti-pastoral system is a good alternative to shifting cultivation. Here the top steep hills are used for forest trees, followed by horticulture, and the lower portion is used for agriculture. Local resources-based soil and water conservation measures like contour bund, bench terraces, half-moon terraces, and grass ways are adopted. Watershed-based farming system thus ensures adequate protection of land against soil erosion, retains maximum rainwater without affecting the crop, and the harvested water can be reutilized for pisciculture or any other purpose according to local used.

10.4 Impact of Watershed Program on Sustainable Agriculture

The input of these program has been beyond expectations. The watershed management program now in operation in different parts of the country has achieved surprising degree of success. The programs are (1) watershed approach for soil and H_2O conservation, (2) DPAP, (3) DF, (3) river valley projects, (5) flood control, and (6) hill area development. At places where watershed have been in operation, (1) soil and H_2O runoff has been checked, (2) denuded forests and devastated pasture lands have been rejuvenated, (3) barren and wastelands have been brought under cultivation, (4) groundwater level and (5) crop yields have gone up, and so have the (6) farmer's income. The impact and effectiveness of watershed initiatives such as Sukhomajri and Bunga have clearly shown that it is possible to embed ecologically sound rainfed agriculture in economically sustainable development. The other good examples are made at Mir (Udhampur, J & K); Alagia Pandiapuram (Tirunelveli, TN), Nartora (Raipur, MP), Jhanwar (Jodhpur), Gursutinala (Palamau, Bihar), Padal Singhi (Beed, Maharashtra), and Dapoli (Maharashtra). As a result of water resources development by the construction of an earthen dam and associated soil conservation and efficient use of irrigation water and dry farming practices besides improvement in animal health, the total food grain production increased from 20 tons to 151 tons, fodder from 56 to 1095 t/year, and milk from 1231 to 3406 liter/day. The entire cost (36 lacs) of the project was recovered in 3 years, and subsequently the watershed society could generate about 1.0 lakh rupee/year.

10.5 Steps Involved in Integrated Watershed Development

1. Crop production assumes importance in watersheds. It is essential to practice in situ moisture consummation and water harvesting.
2. In crop production, priority is for intercropping.

3. Crop substitution, depending upon the suitability of soil, can ensure higher returns.
4. Site-specific crop production techniques have to be adopted.
5. The afforestation component has to be planned carefully and implemented vigorously so as to provide the farmers with affordable access to fuel, timber, and wood; cultivable waste and marginal lands should be used for this purpose.
6. Dryland horticulture will help minimize risk and offer scope to improve returns particularly from marginal lands.
7. Discourage rice cultivation under existing irrigation facilities and use the water elsewhere.
8. More efficient exploitation and use of ground H_2O, coupled with conjunctive use of rain H_2O, and using it for high-value horticultural crops.
9. More efficient use of irrigation water under all resources of water supply.
10. Encouraging other components of the farming system, e.g., dairying.
11. Stall-fed goat and sheep farming needs to be popularized.
12. Sericulture and beekeeping had to be intensified.
13. Processing farm horticultural produce into value-added products offers scope for off-season occupation besides ensuring higher returns.
14. Custom hire services at the village level such as plowing, plant protection, harvesting, etc., will provide employment.
15. The extension and training program should equip farmers to respond to the emerging complexities of agriculture.

11 Plant Ideotype, Their Types, and Ideotype for Dryland Farming

Abstract

An ideotype is a biological model which is expected to perform or behave in a predictable manner within a defined environment, and the selection of a crop ideotype is an important prerequisite for water-saving cultivation and high-yield breeding programs in dryland crops. There are various ideotype types viz., isolation ideotype, competition ideotype, and crop ideotype. The ideotype for dryland farming has the following characteristics: drought tolerance, short growth duration, effective root system, high yield potentiality with altered morphology, contrary to vertically disposed leaves horizontally disposed for better light interception most effective under irrigated conditions.

Keywords

Dryland crops · High yield · Ideotype · Ideotype types · Water saving

11.1 Ideotype

In this chapter, the term "ideotype," which literally means "a form denoting an idea," is suggested for biological models. Ideotypes are biological models that, in the broadest sense, are anticipated to function or behave predictably in predetermined conditions. A crop ideotype is, more particularly, a plant model that, when produced as a cultivar, is anticipated to produce more grain, oil, or other useful products, either in quantity or quality, than other plant models. The fundamental goal of the crop ideotype as a plant model in a water-scarce environment is to produce a large amount of grain or another useful commodity.

In order to cultivate dryland crops in a water-efficient manner and to develop high-yielding varieties, crop ideotype selection is a crucial necessity. Typically, it is based on "defect elimination" or "selection."

S. S. Walia et al., *Rainfed Agriculture and Watershed Management*,
https://doi.org/10.1007/978-981-99-8425-1_11

11.2 Types of Ideotype

- Isolation ideotype
- Competition ideotype
- Crop ideotype
- **Isolation ideotype:** When the plants are space-planted, it is the model plant type that succeeds best.
- **Competition ideotype:** In genetically diverse populations, this ideotype functions well. This idealized plant type for cereals is a tall, green, free-tillering plant that can shade its less aggressive neighbors. The following characteristics of an ideotype for annual seed crops would be present: annual habit, tallness, leafy canopy, tillering or branching, seed size, germination speed, and root traits.
- **Crop ideotype:** Due to its weak competition, this ideotype thrives in commercial crop intensities. A crop ideotype for cereals is an upright, scarcely tilled plant with small, upright leaves.
- **Ideotype for Dryland Farming**
- Drought tolerance
- Short growth duration
- Effective root system
- High yield potentiality with changed morphology, viz. plant with few leaves sufficient to maintain photosynthetic output and growth (to minimize the use of water).
- Contrary to vertically disposed, leaves horizontally disposed for best light absorption most effective under irrigated conditions.

Several other ideotype are:

- **Market ideotype:** Consists of characteristics such as seed size, seed color, baking and cooking quality, etc.
- **Climatic ideotype:** Includes characteristics that are crucial for climatic adaptation, such as resilience to heat and cold, length of maturity, and resistance to drought.
- **Edaphic ideotype:** Consists of resistance to salinity, mineral toxicity, or deficiency, among other things.
- **Stress ideotype:** Demonstrates resilience to biotic and abiotic stress.

11.3 Ideotype-influencing Factors

- **Crop species:** Although branching is significant in dicots, tillering is significant in monocots.
- **Cultivation:** Irrigated crops have characteristics that set them apart from rainfed crops.
- **Socio-economic condition of farmers:** While dwarf sorghum is perfect for American mechanical harvesting, African farmers cannot use it.

11.4 Merits and Demerits of Ideotypes

Advantages: It takes advantage of morphological and physiological diversity. It offers a solution to numerous issues at once and is a productive way to create cultivars for a particular condition or habitat.

Disadvantages: It takes a long time to develop cultivars through ideotype breeding. Individual features that increase yield either generally or in specific genetic and environmental circumstances have not been able to be identified.

11.5 Plant Ideotype for Selected Important Dryland Crops

1. For drought tolerance, a conceptual model in wheat
   (a) Coleoptile length and seed size improve
   (b) Early establishment of crop plant
   (c) Pre-anthesis and early covering of ground
   (d) Biomass (minimize soil moisture evaporation)
   (e) Remobilization/stem reserves and accelerated photosynthesis
   (f) Osmotic balancing
   (g) Stomatal conductance
   (h) Abscisic acid accumulation
   (i) Leaf anatomical traits rolling, waxiness pubescence, and thickness
   (j) High tiller survival and stay-green
2. Cotton (Given by Singh and Narayanan 1993)
   (a) Earliness (150–165 days)
   (b) Less dense, tiny leaves
   (c) Compact and short stature, indeterminate habit
   (d) Sparse hairiness
   (e) Medium to big boll size
   (f) Synchronous bolling
   (g) High reaction to nutrients
   (h) Insect and disease resistance
3. For conditions of high temperature and low moisture, an ideotype of maize
   (a) Change in the site of grain production
   (b) Terminal ear
   (c) Sex determination correlation between genes, environment, and hormones
   (d) Both ear and tassel florets are bisexual and break apical dominance
   (e) Decrease the dependence on favorable water conditions at flowering
4. Ideotype for gram
   (a) Before the starting of reproductive stage, the vegetative growth must be stopped
   (b) To take advantage of a long photoperiod and a comfortable temperature when flowers are in bloom
   (c) Plant is to have erect branching

Table 11.1 Ideotype traits for identification of drought tolerance in certain cereals and legumes

Crop	Seedling establishment	Vegetative	Pre-anthesis	Post-anthesis
Maize	High seedling vigor; high establishment count	High relative stem-and-leaf extension ratio; reduced leaf senescence under severe stress	High moisture content of grain and good seed set; reduced days to anthesis reduced anthesis-silking interval	High yield of grain
Grain legumes	High establishment, high seedling vigor	Short fruit development period; reduced leaf area; reduced stomatal conductance	Closure of stomata; reduced days to anthesis	Good seed size; flower retention capacity
Rice, barley, and wheat	High seedling vigor; high establishment count	Possession of awns, high tiller count (Wheat/Barley)	Reduce days to anthesis	Good seed set/seed size

For identification of drought tolerance in certain cereal and legume crops, ideotype traits are shown in Table 11.1.

Adapted from: (Singh and Narayanan 1993).

Reference

Singh P, Narayanan SS (1993) Biometrical techniques in plant breeding. Kalyani Publishers, New Delhi

12 Introduction, Types, and History of Rainfed Agriculture and Watershed Management in India

Abstract

Rainfed agriculture is a type of farming that relies on rainfall for water. It provides much of the food consumed by poor communities in developing countries. Out of 143 m ha cultivated land about 43 m ha are under irrigation and the rest of the area (about 70%) is rainfed, but all is not too dry and is the food basket for the poor, with a millet-dominant crop pattern. Rain-dependent areas can be broadly split into two: "dryland areas," which receive less than 750 mm (75 cm) of rain a year; and rainfed areas, which receive more than 750 mm (75 cm). On the basis of yearly rainfall, dryland agriculture can be divided into three classes: rainfed farming, dryland farming, and dry farming.

Keywords

Dryland agriculture · Dryland areas · Rainfed agriculture · Rainfall · Rainfed areas

12.1 Introduction

A dryland area is one that receives rainfall between 375 and 1125 mm and has very few irrigation facilities, according to India's fourth five-year plan. Dryland agriculture, as defined by Reddy and Reddy, is the practice of growing crops only when there is sufficient rainfall.

Agriculture that uses rain as its primary source of water is known as rainfed agriculture. It provides a large portion of the food consumed by underprivileged groups in developing nations. For instance, in sub-Saharan Africa, 90% of the land is farmed by rainfed agriculture, whereas 75% of the land is farmed in the Near East and North Africa, 65% of the land is farmed in East Asia, and 60% of the land is farmed in South Asia. The less desirable rainfed regions are home to one-third of the population in the developing countries. Rainfed agriculture is the primary source of

S. S. Walia et al., *Rainfed Agriculture and Watershed Management*, https://doi.org/10.1007/978-981-99-8425-1_12

income for almost 70% of Indians. Because of the regions' high levels of productivity brought on by the green revolution, it therefore offers hope for the future of food security.

Large portions of India are dependent on rain, and they contribute significantly to agricultural output, with just 1/40th of the world land. The second largest number (after Africa) of drought victims of the world live in India. Out of 143 m hectare of cultivated land about 43 m ha are under irrigation and the rest of the area (about 70%) is rainfed, but all is not too dry and is the food basket for the underprivileged, with a crop pattern dominated by millet. Nearly 90% of it located in the north-west part, out of which 60% is located in Rajasthan and sustains a human population of 20 million and a total livestock population of 23 million. The dryland areas contribute about 75% of pulses and oilseeds of the nations' total production and 42% of food grains, almost all the coarse grains. About two-third of rice and rapeseed mustard and one-third of wheat are grown in rainfed areas. A large portion of industrially crucial crops such as cotton, groundnut, and caster are cultivated under dryland conditions and have a great contribution to the production of food, fiber, fuel, furniture timber, etc.

In general, rain-dependent regions can be divided into two categories: "drylands," which receive less than 750 mm (75 cm) of rain annually, and "rainfed areas," which receive more than 750 mm (75 cm). Drylands, which consist of arid and semi-arid habitats, extend from Rajasthan to the southern tip of India as well as from Gujarat in the west to eastern Madhya Pradesh. India has a wide variety of rainfed regions, from those with abundant resources to those with limited resources. Some of the regions with abundant resources are quite productive and have seen a big acceptance of technology. The majority of the places are, however, dry and resource constrained. Farming is a means of survival rather than a growth-oriented endeavor in dry, resource-constrained locations. It is possible to undertake rainfed agriculture across a wide range of soil types, agroclimatic conditions, and annual rainfall ranges between 400 and 1600 mm.

In irrigated agriculture, grain yield has already reached a plateau as a result of issues with nutrient depletion, salinity buildup, and rising water tables. Therefore, the difficulties of the current millennium would be to increase dryland production while guaranteeing resource conservation. Therefore, new approaches would need to be developed to increase the sustainability and productivity of the delicate dryland ecosystems. We must use the most recent technical advancements to turn drylands (gray regions) into green spaces in order to obtain the evergreen revolution.

For the establishment of horti-silvi-pasture, horti-forestry, social forestry, and other similar systems that provide fuel, food, and fodder for livestock as well as an appropriate vegetative cover for ecological management, drylands provide excellent opportunities.

12.2 Area Under Rainfed and Irrigated Conditions in India

Total arable land: 143.8 mha

Dryland: 34.5 mha
Rainfed: 65.5 mha
Irrigated: 43.8 mha

12.3 Types of Dryland Agriculture

Dryland agriculture can be categorized into three groups based on the amount of rain that falls each year.

(a) **Rainfed farming:** Crop cultivation in areas with greater than 1150 mm of annual rainfall. Due to dry spells, crop failures are less likely. There is enough rainfall, but the main issue with rainfed agricultural is drainage. This farming is carried out in humid climates having a period of growth greater than 120 days. Intercropping/double cropping system is suggested in these regions.
(b) **Dryland farming:** Dryland farming refers to the practice of growing crops in areas having annual rainfall greater than 750 mm but less than 1150 mm. While there may be dry spells, crop failures are less common. The main cause of the moisture deficit in these places is higher evapotranspiration than the sum of the precipitation. For dryland farming to be successful in semi-arid areas, soil and moisture conservation methods are essential. The period of growth is between 75 and 120 days. Single-crop/multi-cropping system is suggested in these regions. In black soils or vertisols, drainage facility may be required.
(c) **Dry farming:** Practice of crops grown in regions having yearly rainfall lower than 750 mm and failure of crop caused by protracted dry spells during the crop cycle. In arid areas, dry farming is carried out with the aid of moisture conservation techniques. The period of growth is less than 75 days. Alternate land use system is suggested in these regions. Wind erosion is the main problem in dry farming.

12.4 Indian History of Rainfed Farming and Watershed Management

Period Before Independence

The cultivation of millets, a drought-resistant crop, for food and fodder has been the primary form of agriculture in the dryland regions of India since the beginning of time. The difficulties faced by farmers appear to have been lessened in years with abundant rainfall since there was an abundance of grain and feed. However, crop failure due to insufficient or unpredictable rainfall frequently led to recurrent food shortages and famines because water is the single most crucial ingredient in crop productivity. Drought was a common occurrence. Dryland cultivators' financial situation is highly unstable and tough due to these elements. The First Famine Commission was established by the Indian government in 1880 to address these

other things. Despite this, food shortages have persisted, and gradually, by 1966, food grain imports amounted 10 million tonnes. In the early to late 1960s, Indian agricultural scientists worked with collaborators from around the world to develop high-yielding cultivars and important crops' hybrids, including rice, wheat, sorghum, maize, and pearl millet.

Green Revolution

A breakthrough in productivity and production, primarily of wheat, was brought about by the HYVs program in a brief span of time (wheat revolution, 1968). The period known as the "Green Revolution" was marked by a rise in food production that outpaced population increase (Katyal 1995; Paul 1997; Vishnumurthy 1999).

A benefit came from the Green Revolution in the middle of the 1960s. However, this resulted in a worrying divide in production between rainfed and irrigated agriculture. In order to decrease the detrimental effects of drought on the production of the recently introduced short-duration hybrids of sorghum (CSH-1) and pearl millet (HB-1) in rainfed areas, an extensive research program was seriously considered due to the socio-economic imbalance. The issues with dryland research were made even worse by the droughts of 1965 and 1966.

The ICAR created an extensive dryland agricultural research program with these ongoing issues in mind. As a result, an agreement for bilateral cooperation between the governments of India and Canada saw the beginning of the All India Coordinated Research Project for Dryland Agriculture (AICRPDA) in 1970, with funding from the Canadian International Development Agency. This partnership ran until 1987. This project's singular quality was its emphasis on an approach to identifying and studying the factors that reduce the yield of crop in large semi-arid areas and areas that experience periodical dry spells. The activities of AICRPDA on 23 cooperating centers (now 25) were spread out across the country's various climatic and soil circumstances. The use of good sowing techniques, better varieties, fertilizer use, water and soil conservation measures, weed control, and research efforts allowed for a twofold escalation in the dryland crops yield. The International Crops Research Institute for the Semi-Arid Tropics (ICRISAT) was founded in 1972 by the Consultative Group on International Agricultural Research and is located in Hyderabad.

The Krishi Vigyan Kendras (KVKs) were established in 1977 to show farmers in most areas how to use tested technologies in their crops. With the founding of the All India Coordinated Research Project on Agro-Meteorology (AICRPAM) in 1983 at Hyderabad with 12 cooperating centers (now 25), the dryland agricultural research was boosted even more. In 1985, at Hyderabad, full-fledged research body, the Central Research Institute for Dryland Agriculture (CRIDA), was established as a result of the AICRPDA's modest beginnings. This institute's primary goal was to concentrate on dryland agriculture lead research, leaving local problems and their resolutions to AICRPAM and AICRPDA.

In order to improve infrastructure and develop the zonal research stations under SAUs for undertaking on-site research, the ICAR established the National

Agricultural Research Project in 1987 as a means of reducing regional imbalances in agriculture. A faster growth rate in agricultural production had short-lived effects. The yield and output of numerous crops in the nation have stagnated or decreased almost since the 1990s. The Government of India has been launching a number of programs to reverse the trend of diminishing agricultural production in order to address them.

Following is a timeline of key developments in India's dryland agricultural research

Year	Chronology of events
1920	Scarcity tract development given importance by the Royal Commission on Agriculture
1923	Establishing Dryland Research Station at Manjri (Pune)
1933	Research Stations established at Bijapur and Solapur
1934	Research Stations established at Hagari and Raichur
1935	Research Station established at Rohtak (Punjab)
1944	Monograph on dry farming in India by N.V. Kanitkar, Bombay, Hyderabad
1953	Establishing Central Soil Conservation Board
1954	Establishing Soil and water conservation research institute at Dehradun, with Soil Conservation Centers
1955	Demonstration centers to showcase dry farming practices were started
1959	Central Arid Zone Research Institute (CAZRI), Jodhpur
1970	Research Centers established under AICRPDA in 23 locations
1972	Establishment of ICRISAT, Hyderabad
1976	Establishment of Dryland Operational Research Projects
1977	Krishi Vigyan Kendra was established at Hayathnagar
1983	Starting of 47 model watersheds under ICAR
1984	Dryland Development Board was established in Karnataka
1985	Central Research Institute for Dryland Agriculture (CRIDA) at Hyderabad
1986	NWDPRA programs were started in 15 states by Govt. of India
1988	National Research Centre for Agroforestry National Land Use Policy
1989	Integrated Watershed Development Projects (IWDP) by NWDP
1990	National Watershed Development Project for Rainfed Agriculture (NWDPRA)
1998	National Agricultural Technology Project (NATP)
2005	National Agricultural Innovation Project (NAIP)
2006	National Rainfed Area Authority (NRAA)

References

Kanitkar NV, Sirur SS, Gokhale DH (1968) Dry farming in India. Indian Council of Agricultural Research, New Delhi, p 470

Katyal JC (1995) Rainfed Agriculture Research and Development in India—retrospective and prospective. Workshop on Sustainable Rainfed Agriculture Project, CRIDA, Hyderabad, pp. 49

Randhawa MS (1983) A history of agriculture in India, vol I–III. Indian Council of Agricultural Research, New Delhi

Paul DK (1997) Rainfed farming system development in India : retrospect and Prospect. In: Katyal JC, Farington J (eds) Research for Rainfed farming. CRIDA, Hyderabad, pp 20–28

Vishnumurthy T (1999) Impact of operational research projects on diffusion of dryland technologies. In: Singh HP et al (eds) Fifty years of dryland agricultural research in India. CRIDA, Hyderabad, pp 605–610

Problems of Rainfed Agriculture in India 13

Abstract

The low productivity of agriculture in dry farming regions is due to the cumulative effect of many constraints on crop production. The constraints can be broadly grouped into (a) climatic constraints like rainfall, potential evapotranspiration, temperature, sunshine, wind velocity, and relative humidity; (b) edaphic constraints like physical, chemical, and biological; (c) technological constraints; and (d) socio-economic constraints. There are different types of production constraints according to different crops.

Keywords

Constraints · Crop production · Dry farming regions · Productivity

13.1 Constraints of Rainfed Agriculture

In the dry and semi-arid regions, rainfed conditions continue to be the norm for most cropping. The vast majority of farmers are small-scale operators with little means. Only low-input, unpredictable crop yield subsistence farming is allowed due to the inadequate resource basis. Due to the cumulative impact of numerous crop production restrictions, agriculture in dry farming zones has a low productivity. The following categories best describe the constraints.

S. S. Walia et al., *Rainfed Agriculture and Watershed Management*, https://doi.org/10.1007/978-981-99-8425-1_13

Climatic Constraints

1. **Rainfall:** Scarcity of rainwater which is low and unpredictable with respect to intensity and distribution. Rainfall comes in at least four different varieties.
   (a) The start of the rains may come relatively early or be somewhat delayed.
   (b) Prolonged dry spell in between two rainfall events.
   (c) The rains may end much earlier or last longer than expected.
   (d) Uneven rainfall distribution.
2. **Potential evapotranspiration (PET):** It varies widely due to wide variation in topography, temperature, wind speed, solar radiation, relative humidity, and altitude. The average PET of India is about 1771 mm, which ranges from 1239 mm in the north (Jammu and Kashmir) to 2052 mm in the south-central region (Andhra Pradesh) of the country.
3. **Temperature:** Temperature extremes limit crop growth and productivity. May and June are the hottest months, and the temperature is as high as 48 °C. Heat waves occur during this period.
   In the winter, north-east India endures extremely low temperatures, which in the Ladakh area (cold arid zone) range from −9 to −14 °C to 3–10 °C in arid regions of Punjab, Haryana, and Rajasthan. Frost often occurs in north-west India.
4. **Sunshine:** The daily average sunshine hours are 7–9 in winter over arid zone except cold arid parts, where the duration is less than 2.5 hours per day. During the summer season the arid zones record 9–10 hours per day and highest over Rajasthan, Gujarat, Maharashtra regions, while in cold arid zone, it is less than 8 hours.
5. **Wind velocity:** High wind velocity with desiccating winds causing a high rate of ET and wind erosion. The impact of temperature fluctuations on plant growth is frequently amplified by wind. Blowing sand can be a severe issue in some regions since it can physically harm plants.
6. **Relative humidity:** There is low relative humidity in rainfed areas.
7. Extensive climatic hazards such as drought, flood, weather aberrations, gale, frost, cyclones, and burning winds due to dry, deserted, and denuded situations.

Edaphic Constraints

Physical

1. Uneven topography with high erodibility.
2. Extreme permeability, either shallow or very deep in depth.
3. Difficulty in workability, particularly in vertisols; have high water-holding capacity but are difficult to manage. When dry, these are very rigid but incredibly soft when wet.
4. Soil and sand movement.
5. High surface crusting that damages roots mechanically and results in poor crop stand and high evaporation rates.

6. Soil compaction by tillage implements. Such conditions lead to large amount of water runoff, poor biomass production, and hence low organic matter addition to soil.
7. Poor soil and water conservation are caused by flooding, water logging, and minor field bunds being broken.

Chemical

1. Terrible, marginal territory with unproductive, low-fertility soils.
2. The groundwater contains dissolved harmful salts.
3. Deficiency of major and minor nutrients. For example, poor P availability restricts root development.
4. High concentrations of soluble salts in the surface soils and problematic soils in terms of pH.

Biological Constraints

1. Dryland soils typically have substantially lower levels of biological activity than wet areas due to low organic matter levels and lack of moisture.
2. Organic matter in dryland soils is less chemically and physiologically stable. As a result, the amount of organic wastes is decreased because the soil becomes more acidic, the amount of water and nutrients retained is low, and plant development is poor.

Technological Constraints

1. Choice for crop varieties is limited, which matches the short moist period suitable for cropping as about 60% of the area has 2–4.5 wet months and about 19% of the area has less than 2 wet months per year.
2. Difficulty in designing a suitable cropping system, crop mixture, and crop geometry considering soil type and other flexible components such as the moisture availability index and the duration of the monsoon.
3. Difficulty in evolving static and adaptable technologies fitting into the prevailing as well as anticipated rainfall conditions on a regional basis. The use of high-input-based technology is risky.
4. Particularly on a farm level, difficulty in producing and multiplying seeds.
5. Farming is mostly powered by human and animal labor, which is labor intensive and land saving but expensive and time consuming, making it impossible to cover the entire fertile area during the season's small window of favorable weather.
6. There is very little scope for its improvement due to the low performance and poor availability of biofertilizers and the limited usage of fertilizers.
7. The usage of nutrients and residual moisture is restricted.
8. Difficulty to conserve moisture for proper and timely usage by crops.
9. Difficulty in reclaiming problem soils.
10. Lack of inexpensive energy sources and quick, light tools.

11. Profuse weed, pest, pathogen, and parasite infestation.
12. Unpredicted heavy showers resulting in the accumulation of more than optimum moisture in the soil, impairing field operations including sowing, interculture and harvesting in time.
13. Land improvements with a constrained scope.
14. Water recycling and harvesting provide a high initial investment proposal.
15. Water damage and loss of runoff (up to 50% of incident rainfall) causing loss of soil (greater than 10 t/ha) and formation of gullies resulting in degradation and unsuitability for arable cropping.
16. Poor land capacity, and the scope to improve it is very restricted.

Socio-economic Constraints

1. Peculiar ecological and socio-economic settings.
2. Unstable production and frequent failure of crops rendering farmers poorer.
3. Low cropping intensity, low farm income, malnutrition, and poor quality of drinking water.
4. Small size of farm holdings and high population pressure on land (1.7 persons/ha).
5. Poor quality of produce fetching a lower price.
6. Lack of marketing facilities and market incentives.
7. Low level of literacy and poor resource base of the farmers, though they are economic minded and rational.
8. Being unemployed for the majority of the year.
9. Farmers are to depend on the favors of the monsoon or on financing institutions, which are reluctant to provide assistance as there is more risk in the recovery of the released amount from farming.
10. Farmers of these regions are deprived of innovations, initiative, inspiration, aspiration, and appropriate incentives and appreciations, and they wait, watch, and worry over their future.

13.2 Crop-Wise Production Constraints

Production Constraints in Bajra

Use of Lower Seed Rates and Untreated Seeds

Farmers must be educated by laying down demonstrations on their fields and analyzing the additional costs and additional grains from recommended levels of seed rates as well as treated seeds.

Non-adoption of Chemical Fertilizers Including the Micronutrients

By laying down demonstrations on farmers' fields, they must be educated about the use of adequate fertilizers under the rainfed condition.

Thinning and Gap Filling

The technical feasibility and economic benefits of these practices must be demonstrated to farmers by conducting demonstration trials on their fields.

Non-adoption of Sowing by Ridger Seeder

Tractor-drawn ridger seeder is very heavy and costly, and there is no equally efficient and good ridger seeder drawn by bullocks.

Weed Control

To ease the problem of labor shortages, efficiency should be increased through the use of more efficient implements, tools, and simple machines; such tools must be developed/popularized on a priority basis.

Plant Protection

Suitable varieties resistant to major insects, pests, and diseases must be evolved. Less costly and adequately effective insecticides/pesticides, weedicides, and other plant protection chemicals must be evolved for popularization among farmers.

Lack of Varieties for Dry Farming Conditions

There is strong need for developing suitable HYV highly tolerant to moisture stress conditions.

Soil Moisture

Under the dryland farming situations most of the crops suffer from moisture stress. Appropriate technology for conserving soil moisture as well as other low-cost water harvesting technologies must be evolved and popularized.

Production Constraints in Guar

Use of Improved Varieties

Farmers do not have knowledge about improved varieties that have substantially higher yield potential than the local ones.

Use of Rhizobium Culture

Its utility is shown on paper, but most of the farmers are not convinced about it. It needs to be demonstrated on the farmer's field to convince them.

No/Negligible Use of Fertilizers

The significance of the balance use of fertilizers on yield and profitability of guar needs to be demonstrated by conducting trials on the farmer's fields.

Interculture and Plant Protection Operations Are Not Common

The effect of interculture and plant protection operations on yield and profit needs to be demonstrated by the extension agencies to convince the farmers.

Production Constraints in Wheat

Lower Doses of Fertilizer Application

By performing tests and demonstrations on a farmer's field the economic benefits, particularly the additional grains to the farmers due to the use of full recommended doses of fertilizers, need to be demonstrated to convince them.

Irrigation at Right Time

Farmers are well convinced about the timely irrigation of wheat, and their constraints pertaining to the non-availability of adequate electricity/canal water should be solved by the Govt. to the maximum possible extent.

Effective Weed Control Measures

Farmers are not fully convinced about the magnitudes or increase in yields/profit by applying adequate weed control measures/interculture operations in wheat.

Non-adoption of Various Plant Protection Measures Recommended by the Technical Scientists

Most of chemicals recommended for plant protection measures are very costly. Research efforts are needed to find out cheaper and equally effective chemicals for this purpose. Also, most of the chemicals are very costly. Farmers need to be convinced about recommendations concerning the plant protection measures.

Non-adoption of Weed Control Measures by Majority of the Farmers

Farmers need to be educated to convince them about substantial economic benefits of controlling weeds in their fields. Demonstrations on the farmer's field must be laid out, and its economics be worked out to convince them.

Production Constraints in Gram

Adoption of Improved Varieties

Relatively superior varieties to common ones are available, and well-planned yield maximization demonstration of these recommended varieties needs to be laid out on the farmer's field.

Use of Rhizobium Culture

To convince the farmers about the importance of inoculation of seed with rhizobium culture for increase in yield.

Seed Treatment for Protection Against Termites

The relevant chemicals must be made easily available to the farmers, and they must be educated for the treatment of seeds before sowing.

Uneven Germination and Poor Plant Population Due to Moisture Stress at Planting Time

The available technology for the conservation of soil moisture must be popularized among the farmers by the extension agencies through laying out field demonstration on the farmer's fields.

Very Low Chemical Fertilizer Utilization

Farmers have apprehension that fertilizer usage under rainfed conditions adversely affects the yield. Also many farmers were not aware of the importance of the phosphatic fertilizers.

Production Constraints in Barley

Non-use of Recommended Varieties

In farmer's opinion whatever improved varieties are available are not significantly superior to the locally common varieties; to convince the farmers about the superiority of recommended varieties, demonstrations be laid out on the farmer's field, and its comparative economics worked out.

Application of Very Low Dose of Chemical Fertilizers

Farmers are not fully convinced about adding fertilizers at the correct suggested rates under their condition of not having appropriate irrigation water.

Interculture and Weed Control Measures

Farmers do not adhere to this practice because they are not quite convinced about its economic advantages. They need to be convinced through laying down demonstrations on their fields and working out its comparative economics.

Seed Treatment Against Seed-Borne Diseases and Termites

Farmers need to be educated about its significant advantages through laying down trials/demonstrations on the farmer's field.

Constraints in Raya

(Rapeseed and Mustard)

Thinning of Crop to Make Optimum Spacing for Plants

Farmers need to be convinced about its economic advantages by conducting trials on their fields.

Application of Very Low Doses of Fertilizers

Field demonstrations must be conducted on typical farm conditions to demonstrate the economic benefits of application of the recommended dose of fertilizers as most of the farmers are not quite convinced about this recommendation.

Interculture and Weeding

Farmers are not very convinced about its substantial economic gains. It needs to be demonstrated.

Insect, Pests, and Disease Control Measures

Most of the farmers lack technical know-how about various recommended plant protection measures. Demonstrations must be conducted on the farmer's field, and its comparative economics worked out to educate them.

Tillage—to improve infiltration, facilitate sowing, and weed control, selection of crops and cultivars including intercropping, time and method of sowing, plant population, mulching, and disease and pest management are important. Addition of desired nutrients in optimum amounts and time is also necessary. It must coincide with moisture availability. Maintaining weed-free conditions is equally important. Fertilizers both organic and inorganic are used together. Agricultural residue and legumes are incorporated into crop systems. Drought-tolerant crops and cultivars are coarse cereals, pulses, oilseeds, and legume fodders.

Rainfed Areas' Typical Climate and Soil Conditions

14

Abstract

In rainfed agriculture, soil quality and weather are two key determinants of crop productivity. In general, quality of rainfed soils is poor. These have very little resilience and buffering capabilities. In arid and semi-arid regions, the soils are over-salted (saline-alkali soils), while in sub-humid and humid regions, the soils are acidic (acid soils). Micronutrients and ameliorants, particularly lime, are inadequate and require periodic supplementation. The soils are not suitable for intense cropping since they are primarily coarse textured, extensively deteriorated, have a low capacity to retain water, and have multiple nutritional deficits. Furthermore, climate is one of the major factors influencing crop growth. In dry farming areas, weather, a component of climate, is crucial to crop planning. Rainfall is one of the many weather factors that is crucial to the success of dry farming. Annual rainfall ranges from 400 to 700 mm. Rainfall that is erratic and unevenly distributed makes rainfed farming challenging. Depression is prevalent in late July and early August. Again, late August and September see plenty of rain. Rainfall completely stops until mid-October. In September, the likelihood of precipitation is better than half the average. In different areas of India, different types of climate, viz., arid climate, semi-arid, sub-humid climate, and pre-humid and humid zones are found.

Keywords

Annual rainfall · Crop productivity · Rainfed agriculture · Soil quality · Weather

S. S. Walia et al., *Rainfed Agriculture and Watershed Management*,
https://doi.org/10.1007/978-981-99-8425-1_14

14.1 Soils

Alfisols (Red soils) are agriculturally important soils in drylands. These soils occupy about 20% of the geographical area. The clay is predominantly kaolinitic in nature and dominated by calcium. The color of this clay ranges from red to reddish brown to yellowish brown. Due to their high clay content (between 35% and 65%), soils display undesirable physical characteristics. The reserve of free limes ranges from 3 to 10%. Dryland soils react from being alkaline to somewhat acidic. Saline, saline-sodic, and sodic soils, which are troublesome, are found in patches in low-lying locations. Soils retain less water and nutrients due to their coarse texture.

In terms of fertility, the soils typically have low levels of available phosphate (10–30 kg/ha), total nitrogen (0.03–0.05%), and organic carbon (0.35–0.5%), as well as high levels of available potassium (300–750 kg/ha). In dryland crops, micronutrient deficits are typically not seen. However, crops like groundnut have responded somewhat to the application of boron on degraded soils. Oilseeds and legumes respond well to phosphatic fertilizers, while cereal crops respond quite well to nitrogen fertilizers.

Soil infiltration occurs at a somewhat modest rate of 0.5–0.9 cm/ha. The soils show contraction when dry and significant volume expansion when damp. Wide, deep fractures that extend into the murrum strata form during the shrinking process in medium-depth soils. The deeper layers (phases) of the soil lose moisture more quickly as a result of the crack development.

According to the plowing techniques, slope, and cropping season, additional soils show varied degrees of erosion. The soils were rated as being moderate to very erodible. Because of this, conserving water and soil is necessary for successful cropping. The amount of water and nutrients that may be used for crop production is constrained by the shallowness of the soil. Since they do not retain soil moisture, soils with a depth of lower than 45 cm are often good for Kharif crops. On such soils, crops mature more quickly due to intermittent wetness caused by regular rainfall from June to August.

Greater than 45-cm-deep soils have a great potential to retain and store moisture. To achieve the ideal range of soil moisture on dryland, the amount of rainfall needed is much significant because the rainfall of early monsoon is so little. Usually, the medium-depth soils lack sufficient moisture for sowing. The fact that the medium deep and deep soils received about 200 mm of rain throughout September is the only reason they are moist enough for Rabi cultivation. Therefore, Rabi farming is common, medium deep, and despoils an area where Rabi crops have been planted.

Clay content and soil depth are the key determinants of the moisture-holding capacity of soil. The amount of moisture held in the soil at different stresses affects how quickly the soil loses moisture, despite the fact that clay content is often above 45% in deep as well as medium deep soil. The failure of crops in dryland agriculture is ultimately due to soil moisture content always being below 15 bar.

14.2 Climate

One of the key elements affecting crop growth is the climate. In dry farming areas, weather, a component of climate, is crucial to crop planning. Rainfall is one of the many weather factors that is crucial to the success of dry farming.

The southwest monsoon brings the majority of rainfall to areas with dry soil. The northeast monsoon follows the southwest monsoon as the primary source of precipitation, supplementing it. In different regions of India, there are four distinct types of rainfall. The rain is irregular, typically sparse, and unevenly dispersed. About one-third of Maharashtra State's total territory is considered to be the draught-prone area. The climate here is typically hot, and potential evaporation, which measures evaporation in excessive precipitation, is categorized as semi-arid, for example, potential evaporation about 1300 mm compared to annual precipitation at Solapur is 7/22 mm. This difference results in a 60% deficiency.

Features of the annual rainfall: It varies between 400 and 700 mm. A constant amount of rainfall cannot be guaranteed due to the variability from year to year. Rainfall that is erratic and unevenly distributed makes rainfed farming challenging. Rain starts in the latter part of June or early July.

Depression is prevalent in late July and early August. Again, late August and September see plenty of rain. Rainfall completely stops until mid-October. In September, the likelihood of precipitation is better than half the average ($P = 0.58$).

Dry spells: This is another aspect of rainfall. Monsoon breaks are typically felt (noticed) around late July and August. A break is defined as a time when there hasn't been more than 15 mm of rain in 2 weeks. Over 50 mm of rain typically falls over the week.

Water availability duration: PE and rainfall influence the water availability. When combined, humid and moist periods offer favorable conditions for crop development.

Wind speed: July and August are months with typically high wind speeds. If the wind speed is greater than 18–20 km/h, such a time frame falls during a dry spell. Therefore, the rate of evapotranspiration is high. In November and December, evaporation rates are at their lowest when velocity is low.

Bright sunshine hours: January and February are often the months with the most bright sunshine. The sky often has more dust particles in April and May. August has the fewest hours of brilliant sunshine. This shows cloudy weather without any precipitation.

Humidity: High humidity months are July and September. It is low from February through May. Less relative humidity is noted during dry spells. Evaporation is more necessary when the temperature is high and the humidity is low.

Temperature: Maximum temperatures in the months of late April and early May exceed 41°C. December is when temperatures are at their lowest: 14–15°C is the lowest weekly minimum temperature. The typical climate is semi-arid, with moderate winters and scorching summers. Jowar suffers significantly during extended cold weather, while crops like wheat and gram do poorly because they need a longer cool period.

Different Climates Exist in Various Parts of India.

- **Arid climate:** Western Rajasthan, Bellary (Karnataka), Saurashtra, Kutch, Tirunelveli (TN), and Anantapur (AP)
- **Semi-arid climate:** The region from Kanyakumari in the south to Punjab in the north, covering practically the whole of the peninsula, Gaya-Jumai area in Bihar, and east of Western Ghats.
- **Sub-humid climate:** Northern parts of Punjab, UP, Haryana, West Bengal, Orissa, Bihar, Vidarbha and northern parts of Andhra Pradesh, MP, and from Chennai to Nagapattinam (TN).
- **Pre-humid and humid zones:** West coast, NE region, and adjoining hills

Soil and Water Conservation Techniques in Rainfed Areas 15

Abstract

Agronomic measures and engineering measures are the techniques for soil and water conservation in rainfed areas. Agronomic practices for soil and water conservation help to improve the rate at which water is absorbed by the soil. These are used when the land slope is less than 2%. Agronomic measures include conservation tillage, deep tillage, conservation farming, contour farming, mulching, growing of cover crops, and strip cropping. However, engineering measures are extremely important in preventing erosion on agricultural land. When agronomic methods alone are not sufficient, they are typically utilized where land slope is greater than 2%. These measures include bunding, terracing, trenching, basin listing, and subsoiling. Moreover, Microcatchments are useful for erosion control and in situ moisture conservation. These are used when the land slope is 2–8%. In situ moisture conservation is the practice of storing rainwater where it falls in order to use it effectively. The following list includes the various in situ moisture conservation techniques that affect how the land is configured: furrows and ridges, tied ridging, broad bed furrows, dead furrows, compartmental bunding, scooping, interplot water harvesting, and Zingg terracing or conservation bench terracing.

Keywords

Agronomic measures · Engineering measures · Erosion control · Microcatchments · Soil and water conservation

15.1 Soil and Water Conservation Techniques in Rainfed Areas

The techniques for conserving water and soil can be broadly divided into two groups: agronomic measures and engineering measures.

S. S. Walia et al., *Rainfed Agriculture and Watershed Management*,
https://doi.org/10.1007/978-981-99-8425-1_15

Agronomic Measures

Agronomic practices for soil and water conservation use contour cultivation, mulches, dense-growing crops, strip cropping, and mixed cropping to improve the rate at which water is absorbed by the soil, delay and lessen overland runoff, and reduce evaporation. When the slope of the land is less than 2%, the following agronomic techniques are applied.

Conservation Tillage

- Conservation tillage can encompass a variety of tillage practices, such as minimal tillage, reduced tillage, direct drill, mulch tillage, no-till, stubble mulch farming, trash farming, strip tillage, and plow plant.
- The basic idea of the suggestions for cropland with sophisticated soil conservation programs is conservation tillage.
- The application is mostly where there is extensive automated cereal production or farming with high levels of automation and adequate rainfall. Subsistence agriculture and crop production with low input levels are less suitable for it.
- The principles—to optimize cover by returning crop residues, prevent inverting the top soil, and use a high crop density of vigorous crops—succeed in all circumstances.
- Another benefit of conservation tillage is that it lessens the need for terraces and other permanent constructions. However, there are a number of drawbacks that make it difficult to use conservation tillage in semi-arid environments.
- Crop wastes may be valuable as cattle fodder.
- Planting through surface mulches is difficult for planters pulled by oxen, but it may not be an issue for planters using hand labor.
- Crop wastes may be valuable as livestock fodder.
- Through surface mulches is difficult, while hand-operated planters might not have any issues.

Deep Tillage

- One of the factors contributing to low yields in semi-arid regions is the restricted moisture that is available to crop roots.
- If the rooting depth is increased, the available moisture will be increased.
- Some crops, but not all, and some soils, but not all, benefit from deep tillage.
- Deep tillage requires draught power, which is often scarce in semi-arid locations.
- Subsoiling or ripping can be advantageous in order to break up a pan that is limiting permeability or to boost the soil's porosity.
- The deep placement of fertilizer can also be employed to promote deep root, but again, this method will be challenging to implement in subsistence farming.

Conservation Farming

- It encompasses any agricultural technique that increases yield, or dependability; lowers labor or fertilizer inputs; or does anything else that contributes to better husbandry of land, which we have identified as the cornerstone of effective conservation of soil.

- It includes strip cropping, crop rotations, mixed cropping, alternate cropping, interplanting, organic mulches, deep planting of varieties, surface mulching, dry seeding, etc.

Contour Farming

The earth is unable to absorb all of the rain that falls during heavy rainstorms. Because of gravity, the extra water runs off the incline. Each furrow acts as a rill, speeding up the flow of water if farming is done up and down the slope. The vast majority of rainwater drains away without penetrating the earth. The top layer of healthy soil is washed away, in addition to seeds and fertilizer for plants. All of this leads to a crop growing sparsely and unevenly.

Contour farming is a straightforward technique that involves cultivating as far down the slope as feasible while maintaining the same level.

Benefits:

1. Contour farming decreases soil erosion by reducing soil runoff in comparison with the up and down cultivation in the major groups of soils in India, viz. alluvial soils, black soils, and deep lateritic soil.
2. Crop yields are increased, and soil fertility is preserved through contour farming.

Mulching

- To improve soil structure, reduce evaporation, boost infiltration, keep soil down, protect soil from blowing and washing away, and finally increase crop yields, surface mulches are used.
- Interculture breaks the surface crust that builds after each downpour, controls weeds, creates soil mulch that is between 5 and 7 cm thick, and works to keep the soil moist.

Growing of Cover Crops

- Compared to regular cultivated crops, legumes generally provide a better cover, protecting cultivated land against erosion.
- Depending on the soil and climatic circumstances, cropping systems and crops will differ from place to place.
- Groundnut, black gram, cowpea, and green gram are among the most popular cover crops.

Strip Cropping

The practice of cultivating alternate strips of crops (maize, bajra, jowar, and cotton) that permit erosion and crops (close-growing crops such as green gram, moth, black gram, groundnut, etc.) that resist erosion in the same field. By doing this, the runoff speed is decreased, and the erosion-prone soil is prevented from being swept away.

- Strip cropping is now widely acknowledged for preventing runoff erosion and preserving soil fertility. Strip cropping makes use of various beneficial farming techniques, such as crop rotation, contour cultivation, appropriate tillage, stubble mulching, cover crops, etc.
- The following types of strip cropping exist.

Contour Strip Cropping

- Contour strip cropping involves cultivating crops that permit erosion in strips of appropriate widths along the contour of slopes, interspersed with strips of erosion-resistant crops that protect the soil.
- This practice reduces the length of slopes, controls runoff water flow, aids in sediment deposition, and enhances the rainfall-absorbing ability of the soil.
- Moreover, the dense foliage of erosion-resistant crops acts as a shield against direct impact of rainfall on the soil surface.
- A recommended approach is to introduce crop rotation within the strip arrangement, alternating between erosion-resistant and non-resistant crops.

The most effective and suitable crops for reducing erosion are groundnut, moth bean, and horse gram.

1. In many instances, the standard seeding rates of leguminous crops—aside from groundnut—do not produce sufficiently dense canopies to stop raindrops from hammering against the surface of the soil. Increase the rate of seed by three times.
2. For grains like jowar and bajra, the most effective width of the contour strips is 21.6 m, while for the intervening legume, it is 7.2 m.

Strip widths as suitable for crops permitting erosion and crops resistant on varying slopes

Slope	Width of erosion-permitting crops	Width of erosion-resistant crops
0.5% and below	45.0 m	9.0
Between 1 and 2%	24.0	6.0
Between 2 and 3%	13.5	4.5

Field Strip Cropping

Field crops are planted, but not precisely on contours, in more or less parallel strips across somewhat regular slopes. This technique works well with soils that have high infiltration rates and regular slopes.

Wind Strip Cropping

It entails planting both short-lived and tall-growing crops, like jowar, bajra, and maize, in alternating straight, long, thin, parallel strips laid out directly across the prevailing wind direction, regardless of the contour.

Permanent or Temporary Buffer Strip Cropping

- In the case of permanent or temporary buffer strip cropping, permanent strips of grasses, legumes, or a mixture of grass and legume are laid either in severely eroded areas or in areas that don't fit into a regular rotation, i.e., steep or highly eroded slopes in fields under contour strip cropping.
- These strips are usually permanently or temporarily planted with perennial legumes, grasses, or shrubs and do not participate in the rotation used in regular strip cropping.

Mixed Cropping

- The practice of cultivating two or more crops concurrently in the same field without a specific row design is called mixed cropping. This is done by mixing their seeds.
- A continuous and better land cover, good defense against the beating action of the rain, nearly total defense against erosion of soil, and the assurance of one or more crops for the farmer are important goals of mixed cropping.

Engineering Measures

The construction of mechanical barriers across the direction of the flow of water to contain the runoff for decreasing soil and water loss is an example of an engineering/ mechanical measure. They are extremely important in preventing erosion on agricultural land. When agronomic methods alone are not sufficient, they are typically utilized where land slope is greater than 2%.

The main objectives of the mechanical techniques for controlling erosion are:

1. Lengthen the period of concentration by stopping runoff and, thus, allowing water infiltration.
2. To break up a long slope into multiple shorter ones in order to slow the runoff and stop erosion.

These measures include the following.

Bunding

A bund is an earthen embankment built to decrease soil erosion and manage flow by shortening the slope.

Contour Bunding

- In dry and semi-arid regions with high infiltration and permeability, contour bunding is the most widely used mechanical strategy for reducing soil erosion and preserving moisture.
- In this method, levels that are parallel to the contour of the land are raised in relatively narrow-based embankments at regular intervals.

- It is widely employed on agricultural land with a slope of up to 6% and in regions with an annual rainfall average of less than 1.25 inches.

Graded Bunding/Channel Terraces

- Water flows through graded channels in graded bunding that are built on the upstream side of the bunds and lead to a safe release on grassy waterways.
- Broad-based or narrow-based graded bunds are also possible. On the bottom margin of the canal, a wide, low embankment is constructed from which the soil is extracted to create a broad-based graded terrace. It is dug out at regular intervals along a descending contour with an appropriate longitudinal slope.
- The canal is used where the average annual rainfall is greater than 600 mm and where the land slope is between 2 and 10%.

Grassed Waterways

- Grass waterways are either created or naturally occurring watercourses that are lined with grasses that are resistant to erosion and used to drain surface water from cropland. They are built in accordance with the terrain's slope.
- For the safe disposal of concentrated runoff and to protect the land from rills and gullies, grassed waterways are used.
- Grass waterways are also employed for farm ponds' emergency spillways, for carrying discharge from contour furrows, diversion channels, or natural runoff.
- The grass suitability was determined by how well it provided shelter, how easily it could be established, and how much forage it produced.
- *Panicum repens* was the best-suited grass, followed by *Brachiaria mutica*, *Cynodon plectostachyus*, *Cynodon dactylon*, and *Paspalum notatum*.

Terracing

Terracing is the term for a slope-spanning system of canals and ridges. This is typically done on a steep incline.

Bench Terracing

- Bench terraces are made by partially cutting and partially filling step-like fields along contours. All erosion risks are avoided because the original slope of the ground has been transformed into flat fields.
- The material dug from the upper half of the terrace is used to fill the bottom section of the terrace, resulting in a vertical drop that can range from 60 to 180 cm, depending on the slope, the soil conditions, as well as the economic width needed for straightforward cultural operations.
- Along the outer perimeter of the terrace, a tiny shoulder bund that is 30 cm tall is also built.
- It is typically used on undulating and steeply sloping (16–33%) ground, and it aids in bringing sloping land into several flat strips to facilitate agriculture.
- Bench terraces may be table-top, sloping outward or inward with or without a slight longitudinal slope, and the soil and subsoil may be somewhat absorbent or

poorly permeable depending on the quantity of rainfall in the tract—medium, poor, or heavy.

1. **Table-top (level):** adopted in areas having mild rainfall about 750 mm
2. **Sloping inward:** adopted in areas having rainfall more than 750 mm
3. **Sloping outward:** adopted in areas with low rainfall that is less than 750 mm

Zingg Terracing

- Zingg terracing is used in locations with high rainfall and 3–10% land slopes.
- It is built in medium-to-deep soils.
- The goals to construct Zingg terracing include:
  - To shorten the slope's height.
  - To collect the runoff from higher locations for the benefits of lower area crops.

Trenching

Trenches are made along the contour to conserve moisture and soil and for afforestation purpose.

- Trench size—60 cm × 48 cm
- Trenches spacing—10–30 m
- Excavated materials are used to partially fill trenches, and the remaining soil is used to create a spoil bank.
- The water that is still in the trenches aids in moisture conservation and is useful for planting and sowing.

Basin-Listing

With the aid of a unique tool called a basin lister, small, interrupted basins are created along the counter. Basin-listing is particularly effective on retentive soils with mild slopes and aids in the retention of rainwater as it falls.

Subsoiling

By enhancing the physical properties of a soil, this technique involves breaking up the dense and impermeable subsurface with a subsoiler. Runoff and soil erosion are both decreased by this operation, which does not entail soil inversion and encourages increased moisture infiltration into the soil. At intervals of 90–180 cm, the subsoiler is moved through the soil at a depth of 30–60 cm. The best in situ method for conserving soil and water is sub-soling, which is done with a tractor-drawn chisel plow at a 2-m horizontal spacing. This method promotes early establishment and pasture improvement. Additionally, it will boost productivity and allow for the quick and efficient coverage of a vast region.

15.2 Micro-catchments for Sloping Lands

It is helpful for preventing erosion and preserving in situ moisture.

Characteristics

- Land slope—2–8%
- About 16 m^2 area surrounded by 15–20 cm dirt wall
- Bund height—30–45 cm
- Soil texture - light to moderate
- With staggered planting, in situ moisture conservation
- Best for agroforestry and dryland horticulture

15.3 Check Dam

- Low weirs are typically built over gullies.
- Constructed on long, narrow streams and gullies that have been eroded by floodwater.
- It lowers velocity, lessens erosive activity.
- The stored water enhances soil moisture in the surrounding area and enables percolation to refuel aquifers.
- The check dams should be spaced apart so that the water spread of one is greater than the water spread of the other.
- Height varies from 1 to 3 m depending on bank height, and length ranges from less than 3–10 m.

Percolation pond: In order to increase ground water recharge.

Characteristics

- Shallow depression is created in lower areas of a naturally occurring or diverted stream route.
- It is preferable to be under a gently sloping stream when a narrow valley already exists.
- Situated in naturally permeable soil.
- Adaptable where 20–30 irrigation wells for groundwater are located within the 800–900 m influence zone.
- For the economics, the minimum capacity may be around 5000 m^3. It also serves as a reservoir for silt detention.

15.4 Broad Beds and Furrows

Within the field boundaries, the broad bed and furrow system has been laid out. The area is leveled, and ridgers pulled by either animals or tractors are used to lay it. Rice is grown in furrows, while maize/sorghum is produced in beds.

Characteristics

- Soil moisture conservation in dryland
- Soil erosion controls
- Acts as a drainage channel during heavy rainy days

15.5 In Situ Soil Moisture Conservation

In situ moisture conservation is the practice of storing rainwater where it falls in order to use it effectively. Various methods can be used to accomplish this. The two fundamental requirements in dry areas are to improve the conditions of the soil surface to promote rainfall infiltration and reduce runoff. Therefore, the shape of the terrain affects how easily water can seep into the soil. The following list includes the various in situ moisture conservation techniques that affect how the land is configured.

Furrows and ridges: The field needs to be shaped into furrows and ridges. Along the slope, ridges with widths of 30–45 cm and heights of 15–20 cm are created. When there is heavy rainfall, the furrows safely direct runoff water and prevent water stagnation. When the intensity of the rainfall is lower, they gather and store water. It is appropriate for deep red and medium-to-deep black soils. It is applicable to crops with wide row spacing, such as cotton, maize, chilies, and tomatoes. It is not appropriate for sandy/gravelly soils, shallow black soils, or shallow red soils. It is not appropriate for crops that are broadcast seeded or those that are sown at a narrower row spacing of less than 30 cm. Since furrows are typically constructed prior to sowing, it is possible to sow by dibbling or planting on its own.

Tied ridging: It is a variation of the above system of ridges and furrows that allows rainwater to accumulate in the furrows and gently soak into the soil profile by connecting or tying the ridges at intervals of 2–3 m along the furrows with a small bund.

Broad bed furrows (BBF): For vertisols or black soils in locations with heavy rainfall (>750 mm), ICRISAT has advised using this technique. Furrows that are 15 cm deep and 60 cm wide separate the beds, which are formed here to be between 90 and 120 cm wide, 15 cm tall, and the appropriate length. Beds will slow down runoff's velocity as it occurs, giving more time for infiltration. A 0.6% gradient can be found in the furrows. Crops are planted in wide beds, and any extra water is drained away by several little furrows that may be linked to farm ponds. Bullock or tractor-drawn tools can be used to create it. The bed former cum seed drill allows for simultaneous BBF production and planting, shortening the time between receiving rain and sowing.

Merits of broad bed furrow method

- It aids in storing moisture
- Removes surplus surface runoff safely without causing erosion.
- Improves drainage infrastructure

- Makes dry seeding easier
- The ability to support a wide range of crop geometries, including both close and wide row spacing
- It works well for both monoculture and intercropping systems.
- Seed drills are used for sowing

Dead furrows: About 20 cm deep furrows are created at a spacing of six to eight rows of crops during or just after sowing. There is no crop grown in the furrow. Before the beginning of the heavy rains (Sep–Oct), between two rows of the crop, dead furrows can also form. Mostly in red soils, it is possible to use a wooden plow. The dead furrows lengthen the window of opportunity for infiltration.

Compartmental bunding: Using a bund formation, tiny bunds of 15 cm width and 15 cm height are formed in both directions to divide the space into little basins or compartments that are square or rectangular and range in size from 6 × 6 m to 10 × 10 m. They are helpful for temporarily storing rainwater, which promotes high infiltration and high soil moisture storage. Black soils with a slope of 0.5–1% are advised. Sorghum, sunflower, and maize all work well for this form of bunding.

Scooping: Rainwater can be retained on the surface of soil for longer periods of time by creating small depressions or basins in the soil surface. By capturing eroding material, they also lessen erosion. According to studies, this method can cut runoff by 50% and soil loss by 3–8 t/ha.

Interplot water harvesting: A tiny catchment's upper half is used for crop cultivation, while a portion of it is used for water extraction. There may be 1:1 cropped: catchment area or 1:2 catchment: cropped area.

Zingg terracing or conservation bench terracing: They were invented in the USA by A. W. Zingg. It is used in regions with low-to-moderate rainfall, especially in black soil. In order to spread the runoff water from the remaining two-thirds of the field, the bottom one-third of the ground near the contour is leveled in this form of land shaping. Even in years with insufficient rainfall, this rainfall multiplication strategy guarantees at least one successful crop in a third of the land. The leveled area (receiving area) is typically where water-intensive crops (like paddy) are grown during medium rainfall years, while the unleveled area (donor area) is where dry crops are grown.

Water Harvesting: Importance, Its Techniques

16

Abstract

Rainwater harvesting, in its broadest sense, is a technology used for collecting and storing rainwater for human use from rooftops, land surfaces, or rock catchments. The rainwater harvesting system incorporates multiple components that are employed at different stages to facilitate the transportation, filtering, and storage of rainwater. The various constituents comprising a rainwater harvesting system include catchments, coarse mesh, gutters, conduits, first-flush, filters, and storage tanks. Water harvesting methods comprise water harvesting in situ which includes contour farming, strip cropping, live beds, vegetative hedges, tied ridging, broad bed and furrows, dead furrows, bunds (contour and graded), compartmental bunds, and terracing. Another method for rainwater harvesting is runoff harvesting, which has again different methods according to arid and semi-arid regions. In arid areas, runoff farming, water spreading, and microcatchments are practiced, whereas in semi-arid areas, dug wells, tanks, minor irrigation tanks, nala bunds and percolation tank, stop dams, and farm ponds are used. Water harvesting techniques are inter-plot water harvesting, inter-row water harvesting, and water harvesting in farm ponds.

Keywords

Arid areas · Components · Rainwater harvesting · Semi-arid areas · Water harvesting methods · Water harvesting techniques

16.1 Importance of Rainwater Harvesting System

The process of gathering, storing, and collecting rainwater into reservoirs or other natural storage areas for later use is known as rainwater harvesting.

S. S. Walia et al., *Rainfed Agriculture and Watershed Management*,
https://doi.org/10.1007/978-981-99-8425-1_16

In its broadest definition, rainwater harvesting is a technology used to collect and store rainwater for human consumption from rooftops, land surfaces, or rock catchments using both low-tech methods like jars and pots and high-tech methods like designed methods. Due to the temporal and geographic diversity of rainfall, rainwater harvesting has been done for more than 4000 years. It is a crucial source of water in many places with heavy rainfall but no established, centralized supply infrastructure. Additionally, it is a suitable choice where there is a scarcity of high-quality fresh surface water or groundwater. Rainwater can be collected and stored effectively, made available, and used in place of low-quality water. The following justifies the need to collect rainwater:

- Rainwater harvesting is better for the ecosystem and the globe since it recharges the groundwater level. The quality of the groundwater is improved. An increase in the water levels in dry bore wells and wells.
- The practice of collecting and harvesting rainwater, and subsequently storing it in tanks, can contribute to personal utility bill reduction by reducing reliance on public water supplies.
- This system will guarantee the consistent availability of water, independent of any unfavorable circumstances such as supply disruptions or water scarcity. An optimal resolution for regions characterized by insufficient water supplies.
- Rainwater has served as a fundamental catalyst for the development and sustenance of the natural environment and ecosystem for an extensive duration spanning millions of years. Consequently, the utilization of this substance is more advantageous for horticultural purposes due to its relative purity and absence of toxic elements.
- By lowering the likelihood of runoff during periods of heavy rainfall, the collected rainwater decreases floods and aids in reducing soil erosion.
- The process of establishing and implementing a rainwater collecting system is characterized by its simplicity and cost-efficiency. Furthermore, it facilitates expedient retrieval of the accumulated water.
- The practice in question functions as a sustainable and environmentally conscious approach that effectively mitigates carbon emissions and advocates for methods that are conducive to the well-being of the environment.
- The lessening of the consequences of drought and achieving drought resistance.
- Energy savings from not having to remove as much groundwater.
- Reduction in the clogging of stormwater drains and flooding of highways.

16.2 Components

The rainwater harvesting system incorporates multiple components that are employed at different stages to facilitate the transportation, filtering, and storage of rainwater. The various constituents comprising a rainwater harvesting system include the following.

16.3 Catchments

The catchment refers to the geographical region that directly absorbs precipitation and serves as the primary source of water for the surrounding hydrological system. A catchment refers to a paved surface, such as a courtyard or terrace, that is typically associated with a residential or commercial structure. Additionally, it may encompass a region such as a spacious expanse or a cultivated grassy area. Rainwater harvesting commonly involves the utilization of roofing structures or surfaces constructed from corrugated sheets, reinforced cement concrete, or galvanized iron.

Coarse Mesh

The coarse mesh serves as a barrier, effectively obstructing the movement of large fragments of waste or other substances from the catchment area into the subsequent system. This measure ensures the prevention of visible pollutants from entering the system and causing water contamination.

Gutters

Gutters are structural components positioned along the perimeter of a catchment or inclined roof, serving as routes for water drainage. The system is responsible for the collection and transportation of rainwater to a designated storage unit or tank. The shapes of these objects might vary, encompassing forms such as semi-circular and rectangular, among others. They can be made using different materials.

Conduits

Conduits refer to the drains or pipelines responsible for transporting rainfall from the rooftop area or catchment to the collecting system. These structures have the potential to be constructed using a wide range of readily accessible materials, such as galvanized iron or polyvinyl chloride. The selection of pipe diameter for installation is contingent upon the magnitude of rainfall intensity.

First-flush

The valve functions to prevent the ingress of initial rainfall water into the system, facilitating its expulsion. This phenomenon occurs when the initial precipitation of rainwater is collected or utilized in some manner.

Filters

Filters (sand and charcoal) are employed for the purpose of eliminating contaminants present in water obtained from the catchment region. The filter unit comprises a chamber containing a combination of filtering components, including gravel, charcoal, coarse sand, and fiber. The process involves the removal of dirt and debris from the water prior to its entry into the storage unit or recharge structure.

Storage Tanks

The storage tanks or units serve as the destination for the collection and storage of rainwater. The objects can exhibit a diverse range of shapes, materials, and sizes, contingent upon the specific needs and specifications. The shapes utilized in this context include rectangular, square, and cylindrical forms. Commonly utilized construction materials include plastic, reinforced cement concrete, and metal sheets, among others. Additionally, their placement can be determined based on the availability of space. Additionally, it is necessary to implement maintenance protocols in advance.

16.4 Methods of Rainwater Harvesting

The term "water harvesting," which covers a variety of techniques for gathering and preserving different types of runoff from diverse sources, is frequently used in this context. Water can be collected "in situ" or in ponds or dugouts. In dryland agriculture, "water harvesting" typically refers to the process of collecting extra runoff in a storage tank and utilizing it to boost crop output there or in other places.

Protective irrigation is carried out using water collected in the farm pond. The water held in other structures will rehydrate the groundwater and be used for irrigation, either protective or supplemental. The practice of gathering and retaining water on the soil's surface for future utilization is commonly referred to as water harvesting. The approach pertains to the facilitation, acquisition, retention, and preservation of indigenous surface runoff for agricultural purposes within arid and semi-arid geographical areas. Water harvesting is used in both arid and semi-arid environments, albeit with distinct variations. In locations characterized by aridity, the proportion of the collecting area, also known as the catchment area, is significantly greater in comparison to the command area. In semi-arid locations, runoff is not caused in the catchment area; instead, only the excess rainfall is collected and stored. However, both dry and semi-arid areas use a variety of water collection techniques. Runoff from cultivated fields should be led through grassed waterways to suitably located excavated or embankment farm ponds. The runoff harvesting, storage, and utilization involve:

(a) Selection of catchment
(b) Selection of site and estimation of size of farm pond
(c) Construction of farm pond including inlet
(d) Laying of silt trap and spillways
(e) Minimization of seepage, percolation, and evaporation losses; utilization of stored water for life-saving irrigation of kharif crops or for the establishment of rabi crops in the catchment

Water Harvesting In Situ

Water harvesting techniques adopted "in situ" are inexpensive to augment rainwater use efficiency in drylands; any process of reducing the runoff volume will increase moisture infiltration in the soil, leading to increased moisture volume.

The implementation of "in situ" water collection and runoff recycling techniques represents effective strategies for preserving crop viability in times of moisture scarcity. The utilization of "in situ" water collection techniques contributes to the enhancement of crop establishment, particularly for small-seeded crops such as pearl millet. This is achieved by alleviating the negative impacts of soil crust formation on germination and addressing the issue of seedling burial caused by rainfall during the germination phase. This approach additionally facilitates the secure disposal of surplus rainwater in regions characterized by high levels of precipitation. Appropriate land configurations for "in situ" moisture conservation are broad beds and furrows, graded borders, strips, and inter-row and inter-plot water harvesting systems. For shallow alfisols and related soils even dead furrows serve this purpose very well. Strategies for "in situ" moisture conservation involve contour farming and installing barriers on the contour.

Contour Farming

Contouring, also known as contour farming, involves conducting all agricultural activities along the contour lines of the land, as opposed to vertically up and down the slope. Contour farming is widely recognized as a highly efficient and cost-effective approach to erosion prevention and moisture conservation. Contour fanning involves the execution of tillage activities in alignment with the contours of the land. It creates numerous ridges and furrows, which increase surface detention capacity, increase the time for water to infiltrate, and decrease the amount of runoff, harvesting a sizable amount of runoff inside the field itself. Contour farming is, however, effective on gentle slopes of up to 5%. Steeper slopes require slope management techniques. Contour farming can be sometimes inconvenient because it possibly involves frequent turning of farm vehicles, and loss of area, which has to be put into buffer strips.

Strip Croppin

Another facet of contour farming involves the use of strip cropping, which entails cultivating crops in elongated and narrow strips that are aligned with the contour of the land. In this agricultural system, open-row crops such as corn are cultivated in alternating strips with close-canopy crops such as soybeans, alfalfa, and others. The close-canopy crop is frequently cultivated in a contour strip positioned downhill from the open-row crop. It is also important to establish buffer strips on the contour. There are various types of strip cropping. On the basis of objectives, vegetative materials used, and field design adopted, buffer strips are called contour strip cropping, buffer strip cropping, barrier strips, border strips, or field strips. In addition to reducing runoff amount, establishing strips against the prevailing wind direction also decreases wind erosion.

Live Beds

One or two live beds, measuring 2–3 m in width, can be established inside the interbund space, either on contour or on grade. The selection of vegetation for the beds is according to the preferences of the cultivator, allowing for the inclusion of annual, perennial, or a combination of both plant species.

Vegetative Hedges

Vegetative hedges are strategically planted along the contour in order to serve as a physical barrier, hence prolonging the duration of water infiltration into the soil. Vegetative hedges are mostly established using grasses of the bench-type variety. Vetiver (*Vetiveria zizanioides*), commonly known as khus grass, is a highly suitable grass species for the establishment of vegetative hedges in tropical eco-regions. Vetiver is a densely tufted bunch of grass, which can be easily established. In addition to grasses, vegetative hedges can also be established from woody perennials, e.g., *Leucaena leucocephala* and *Gliricidia sepium*, and adequately maintained. Runoff velocity, runoff, and soil erosion can be decreased by using vegetative hedges. However, closely spaced and narrow strips of grass or woody perennials are likely to be more effective in reducing runoff and soil erosion than narrow or single-row hedges.

Tied Ridging

This is an altered iteration of the ridge furrow system, wherein ridges are interconnected at regular intervals of 15–20 m to facilitate the gathering of precipitation in the furrows and its subsequent infiltration into the root zone. The system effectively mitigates the loss of sediment and nutrients caused by runoff, in addition to its rainwater conservation capabilities.

Broad Bed and Furrows

The broad bed and furrow system involves the creation of alternating bed and furrow formations. This particular approach is particularly well suited for black soils, wherein crops are planted on pre-established beds. The aforementioned system is established prior to the commencement of each season and is consistently upheld in subsequent years. The act of planting has been completed inside the designated area. Typically, the furrow depth is maintained at a depth of 0.15 m, while the spacing between furrows is consistently set at 1.5 m.

Dead Furrows

Dead furrows are strategically placed over the topography of rolling lands in order to intercept and redirect the flow of runoff water. The spacing between deceased furrows exhibits variability, ranging from 2 to 5 m or 4 to 7 rows of crops. The efficacy of this technique is seen to be high in alfisols.

Bunds (Contour and Graded)

Earthen embankments of reduced height are implemented along the slope of cultivated areas after the determination of waterway placement. The bunds function to intercept runoff, increase infiltration opportunity time, and dispose of excess rainfall safely. Such bunds can be either contour bunds or graded bunds; many a time, these bunds are adjusted with field boundaries if deviation from grade or contour is not too much; no doubt, spacing depends on many engineering considerations.

Bunds particularly are made with the following objectives:

1. To reduce runoff
2. To reduce soil loss
3. To divert runoff

Thus, graded bunds along with waterways and water harvesting structures not only check soil erosion but also provide an ideal rainwater management system for many watershed situations.

Compartmental Bunds

Compartmental bunds, converting the area into square/rectangular parcels, are useful for temporary impounding of water for improving moisture status of the soil. These are made using bund formers. In medium-deep black soil, they are found advantageous in storing the rainfall received during the rainy season in the soil profile, thereby augmenting the soil moisture for use by rabi crops. The size of the compartments may be fixed.

Compartmental bunds are usually practiced in deep black soils with relatively flat terrain and under low-to-moderate rainfall. The field is laid out into compartments of 6 m × 6m to 10 m × 10 m using a bund former. The harvested water in these compartments conveniently infiltrates into the root zone and is conserved in situ.

Terracing

Cultivated areas characterized by land slopes above 10%, particularly in hilly regions, should be subjected to bench terracing techniques, whereby the ground is transformed into a sequence of platforms. The breadth of a bench terrace is contingent upon the gradient of the terrain and the allowable extent of excavation. Bench terracing has proven to be quite successful in mitigating soil erosion in regions characterized by hilly topography. In Ootacamund, the implementation of certain measures has resulted in a significant reduction in the annual erosion rate.

Specifically, the erosion rate on 25% slope soils has decreased from 39 metric tons per hectare to less than 1.0 metric tons per hectare. In locations where there is an abundance of scattered stones, it is possible to utilize loose stone walls as risers in the construction of bench terraces.

Runoff Harvesting

In rainfed areas, one or two occasional heavy rainfall occurs which creates runoff. In vertisols, the runoff is about 60%, and in alfisol it is about 40%. Hence, efforts should be made for proper utilization of the runoff after taking measures for "in situ" soil moisture conservation.

16.5 Methods of Runoff Harvesting

Arid Regions

The catchment region should possess a sufficient water supply to adequately sustain agricultural growth, while the farming methods employed should optimize water utilization. Perennial crops are often considered to be viable options due to their possession of deep-root systems that enable them to utilize runoff water stored at significant depths within the soil, hence minimizing losses resulting from evaporation.

Runoff Farming

The Negev Desert in Israel has ancient runoff fans that encompassed multiple farmed fields, which were supplied with water from watersheds ranging in size from 10 to 50 hectares. The watersheds were partitioned into discrete catchment areas ranging from 1 to 3 hectares in size. This division facilitated the collection of runoff water through the construction of channels on the hillsides. The tiny size of these catchment areas effectively mitigated the risk of excessive water accumulation that could otherwise become unmanageable. The water channels that directed water toward the cultivated fields were constructed in a terraced manner and incorporated stone spillways. This design allowed for the controlled transfer of excess water from higher fields to lower ones. Farmers implemented the construction of diminutive check dams utilizing boulders, strategically placed across little valleys, in order to effectively channel water toward agricultural fields.

Water Spreading

In arid areas, the limited rainfall is received as short, intense stones. Water swiftly drains into gullies and then flows toward the sea. Water is lost to the region, and floods caused by this sudden runoff can be devastating often to areas otherwise untouched by the storm. Water spreading is a simple irrigation method for use in such a situation. Flow waters are deliberately diverted from their natural courses and spread over adjacent plains. The water is diverted or retarded by ditches, dikes, small dams, or brush fences. The wet flood plains or valley floods are thereby used to grow crops.

Microcatchments

A plant can grow in a region with too little rainfall for its survival if a rainwater catchment basin is built around it. Microcatchments used in the Negev desert range from 16 m^2. Each is surrounded by dirt wall of 15.20 cm height. At the lowest point within each microcatchment, a basin is dug about 40 cm deep, and a tree is planted in it. The basin stores the runoff from microcatchment.

Semi-arid Regions

Water harvesting techniques followed in semi-arid areas are numerous and also ancient. The objectives of water harvesting structures are:

1. To store water for supplemental and off-season irrigation
2. To act as silt detention structure
3. To recharge the groundwater
4. To raise aquaculture species
5. Recreation and allied agricultural uses

Thus in most situations, in addition to reviving traditional water harvesting systems, other measures are also needed. Some of such typical water harvesting structures adopted in watershed are as follows.

Dug Wells

Hand dug wells have been used to collect and store underground water, and this water is lifted for irrigation. The quality of water is generally poor due to dissolved salts.

Tanks

Runoff water from hillsides and forests is collected on the plains in tanks. The traditional tank system has following components, viz. catchment area, storage tank, tank bund, sluice, spillway, and command area. The runoff water from catchment area is collected and stored in storage tank on the plains with the help of a bund. To avoid the breaching of tank bund, spillways are provided at one or both the ends of the tank bund to dispose of excess water. The sluice is provided in the central area of the tank bund to allow controlled flow of water into the command area. The command area of many tank ranges from 25 to 100 hectares. In areas receiving annual rainfall of about 1000–1500 mm with runoff around 50%, where canal and well irrigation is not feasible due to topography and underground water table, tanks are found suitable. An earthen dam of 12m in height stores about 55,600 cubic l of water from a forest watershed of 10 hectares. This tank can provide supplemental irrigation for 20 hectares of rainfed farm land. However, there are a few tanks with command area of more than 1000 hectares.

Minor Irrigation Tanks

Minor irrigation tanks are constructed across the major streams with canal system for irrigation purpose by constructing low earthen dams. A narrow gorge should be preferred for making the dam in order to keep the ratio of earthwork to storage as minimum. The height of the dam may vary from 5 to 15 m. The tanks are provided with well-designed regular and emergency spillways for safety against side cutting. In micro-watersheds, water harvesting bundhies, similar to small-scale irrigation tanks are also recommended. By and large, the water harvesting bundhies are not integrated with extensive canal system.

Nala Bunds and Percolation Tank

Nala bunds and percolation tanks are strategically positioned within nalas that possess permeable formations, primarily aiming to facilitate the replenishment of groundwater resources. Percolation tanks offer a further benefit in the form of stringent regulation on the influx of silt load into downstream reservoirs. The percolation tanks encourage the digging up of wells downstream of recharged area for irrigation purposes. The percolation tanks are provided with emergency spillways for safe disposal of flow during floods.

Stop Dams

Stop dams are permanent engineering structures constructed for raising the water level in the nala for the purpose of providing life-saving irrigation during drought periods. These are located over flat nalas at narrow gorges carrying high discharge of long durations. They are created over stable foundation conditions where hard murrum or rocks are encountered. For the stop dam, a site with larger water storage capacity would be desirable.

Farm Ponds

Farm ponds are typical water harvesting structures constructed by raising an embankment across the flow direction or by excavating a pit or a combination of both. Dugout ponds in light soil require lining of the sides as well as the bottom with suitable sealants. This is not the case with heavy black soil. Considering the benefit of the harvested water, different type of linings, viz. brick, concrete, HDPE, etc. may be used.

Types of Farm Pond

Farm ponds are compact reservoirs utilized for the purpose of accumulating and retaining runoff water. Farm ponds can be categorized into three types based on their construction method and their appropriateness for various topographic conditions. The first type is excavated farm ponds, which are suitable for flat terrains. The second type is embankment ponds, which are designed for hilly and rugged areas with frequent wide and deep watercourses. The third type is excavated-cum-embankment ponds, which are recommended for soils that have a mild-to-moderate sloping topography.

Excavated farm ponds are of three types according to their shape. They are:

1. Square
2. Rectangular
3. Circular ponds

Among these, the circular ponds have the geometric advantage that they have the highest storage capacity and have least circumferential length for a given surface area and side slopes. However, their curved shape is disadvantageous in that a substantial area is normally lost for agricultural operations. Hence, either square or rectangular ponds are suggested.

16.6 Selecting a Catchment for a Farm Pond

The careful selection of a catchment area for the placement of a farm pond holds significant importance due to multiple reasons. The selection is based on the possibility for generating a substantial amount of runoff. An excessively large catchment area can lead to the quick accumulation of sediment, whereas an insufficiently small catchment area may result in inadequate water inflow to the pond. The quantity of runoff from various catchments is contingent upon multiple factors, as explained below:

Factors governing the amount of runoff are as follows

Major factor	Associated factors
1. Rainfall	1. Rainfall intensity
	2. Rainfall duration
	3. Rainfall distribution
	4. Events of rainfall causing runoff
2. Land topography	1. Degree of land slope
	2. Length of run
	3. Size and shape of the catchment
	4. Extent of depressions and undulations on the catchment
3. Soil type	1. Soil infiltration rate
	2. Antecedent soil moisture
	3. Soil texture
	4. Soil structure
	5. Soil erodibility characteristics
4. Land use pattern	1. Cultivated or uncultivated or partially cultivated
	2. Under pasture or forests
	3. Bare fallow or with vegetation
	4. Soil and moisture conservation measures adopted or not
	5. Crop cultural practices that are adopted

16.7 Techniques of Water Harvesting

1. **Inter-plot water harvesting:** This approach is applicable in regions characterized by limited precipitation levels (<500 mm) and challenges in achieving the whole growth cycle of a single crop. This strategy involves the cultivation of a designated piece of land while utilizing the rest area for water gathering purposes. Typically, the undeveloped land is subjected to compaction or treatment methods that promote the occurrence of runoff. Surface alteration may be necessary in order to achieve effective runoff. This particular approach is well suited for places characterized by aridity. Runoff can be initiated through several methods, such as the implementation of cover films (composed of plastic or rubber), the creation of a hydrophobic layer (using wax), the compaction of the surface, or the application of sodic soil onto the surface.
2. **Inter-row water harvesting:** In certain regions, insufficient precipitation may provide challenges to crop cultivation. Consequently, implementing water conservation techniques, such as furrow irrigation and cultivating crops within furrows, may potentially enhance agricultural productivity.
3. **Water harvesting in farm ponds:** A fraction of the surplus runoff water, following the implementation of optimal in situ moisture conservation practices, is accumulated in agricultural reservoirs known as farm ponds. Ideally, it is recommended that the pond be situated in the lower sections of the field in order to optimize storage capacity and minimize seepage losses. A farm pond with a capacity of 150 m^3 and side slopes of 1.5:1 is deemed adequate for every hectare of catchment area in black soil. The agricultural ponds can exhibit many shapes, including round, square, or rectangular configurations. Nevertheless, ponds with an earthen or rectangular shape are more practical for the purpose of collecting and utilizing runoff water.

16.8 Advantages of Rainwater Harvesting

- Reduced expenses
- Assists in mitigating water expenses
- The need for water is reduced
- Mitigates the reliance on water imports
- Advocates for the conservation of both water and energy resources
- Enhances the overall quality and volume of groundwater resources
- The utilization of a filter system for landscape irrigation is not necessary
- This particular technology exhibits a rather straightforward installation and operation process
- The implementation of this practice has been found to effectively mitigate soil erosion, stormwater runoff, flooding, and the contamination of surface water by fertilizers, pesticides, metals, and other sediments

- The water in question serves as a highly beneficial resource for the purpose of irrigating landscapes, as it lacks any chemical substances, dissolved salts, and minerals

Practical 17

17.1 Seed Treatment, Seed Germination, and Crop Establishment in Relation to Moisture Content

Seed Treatment

Seed treatment refers to the application of fungicide, insecticide, or a combination of both, to seeds so as to disinfect them from seed-borne or soil-borne pathogenic organisms and storage insects. It also refers to the subjecting of seeds to solar energy exposure, immersion in conditioned water, etc.

The success of dryland agriculture depends on seasonal rains. Several management strategies are followed to enhance the production potentiality of the crop under rainfed condition. Seed hardening is one such seed management technique, where seeds are hydrated and then dried to their original moisture content to tide over the stress environmental conditions at field. Though simple water acts as a good hardening agent, the efficiency could be enhanced by the use of chemicals where selectivity depends on crop.

The seed hardening techniques recommended are as follows.

Crop: Pearl millet

Chemical and concentration: 2% potassium chloride

Methodology: Dissolve 20 g of salt in 1000 mL of water. Soak 1 kg of seed in 650 mL of this solution for 10 h and dry back to original moisture.

Crop: Sorghum

Chemical and concentration: 2% potassium dihydrogen phosphate

Methodology: Dissolve 20 g of salt in 1000 mL of water. Soak 1 kg of seed in 650 mL of the solution for 16 h and dry back to original moisture or weight.

Crop: Cotton

Chemical and concentration: 2% potassium chloride

Methodology: Dissolve 20 g of salt in 1000 mL of water. Soak 1 kg of seed in 650 mL of the solution for 10 h and dry back to original moisture or weight.

Crop: Sunflower

S. S. Walia et al., *Rainfed Agriculture and Watershed Management*,
https://doi.org/10.1007/978-981-99-8425-1_17

Chemical and concentration: 2% potassium chloride

Methodology: Dissolve 20 g of salt in 1000 mL of water. Soak 1 kg of seed in 650 mL of the solution for 12 h and dry back to original moisture or weight.

Crop: Pulses, black gram, green gram

Chemical and concentration: 100 ppm zinc sulfate, 100 ppm manganese sulfate

Methodology: Dissolve 1000 mg of salt in 1000 mL of water. Soak 1 kg of seed in 350 mL of the solution for 3 h and dry back to original moisture or weight.

Cost of treatment: Rs. 0.50–1 kg

The percentage of yield increase: 10–15%

Designer Seed Technology

The designer seed technology has been developed for cotton, black gram, and direct-sown paddy. It is an integrated seed treatment method which involves addition of nutrients and plant protectants to the seed for improved emergence and establishment.

1. **Cotton**
   (a) The acid delinted cotton seeds are mixed with polykote @ 3 g by diluting in 5 mL of water followed by carbendazim @ 2 g/kg + imidacloprid @ 5 g/kg + Pseudomonas 10 g/kg + Azophos 120 g/kg in sequential manner.
   (b) The treated seeds recorded 12% yield increase over control. The cost benefit ratio of this is 1:1.6
2. **Black gram**
   (a) For rainfed ecosystem, the seed treatment viz., hardened seed + polykote @ 3 g + dimethoate @ 5 mL + *Trichoderma viride* @ 4 g + *Rhizobium* @ 20 g + Azophos @ 40 g + coating with 2% Mg CO_3/kg of seed registered higher physiological seed quality attributes and increased yield of 66 kg /ha.
3. **Direct sown rice**
   (a) The rice seeds were fortified with KCl (1%) for 16 h, followed by sequential coating with polykote @ 5 g kg^{-1} + carbendazim @ 2 g kg^{-1} + imidacloprid @ 5 mL kg^{-1} + *Pseudomonas fluorescens* @ 10 g kg^{-1} + Azophos @ 120 g kg^{-1}.

Seed Germination

The activation of metabolic machinery of seed embryo is the first and foremost step to initiate the seed germination process. Thus, seed germination is the process of reactivation of the metabolic activity of the seed embryo, resulting in the emergence of radical (root) and plumule (shoot), thus leading to the production of a seedling or a young plant.

Two Types of Germination Are Commonly Found in Cultivated Plants

1. **Epigeal germination:** Seed germination in dicots in which the cotyledons come above the soil surface. In this type, the hypocotyl elongates and raises the cotyledons above the ground surface; it is called as epigeous or epigeal germination. This type of germination is very common in beans, gourds, castor, tamarind, onion, etc.
2. **Hypogeal germination:** Seed germination in dicots in which the cotyledons remain below the soil surface. In this type, the epicotyl elongates and the hypocotyl does not raise the cotyledons above ground, which is called as hypogeous or hypogeal germination. This type of germination is common in mango, custard apple, pea, gram, lotus, maize, etc.

Moisture and Germination

Moisture availability has been regarded as the major factor determining the onset of germination. Germination is assumed to start if the soil moisture content is higher than 1.2 times wilting point. As long as this condition is satisfied, germination proceeds unhampered through its various phases, until, at the end of 7 days, germination is assumed to be completed and emergence occurs. If the soil dries out to less than 1.2 times wilting point within 4 days after the onset of germination, the process is halted and will resume after rewetting from the point where it stopped. If drying out occurs 4 or more days after the onset of germination, deterioration of the germinating seeds takes place. If the dry conditions persist for more than 6 days, the seeds are assumed dead, and there will be no crop established from that seed.

Crop Establishment

In soils with low moisture content, the germination rate will be lower and emergence will be slower. As the seedling develops it becomes more susceptible to low soil water. This can vary with different soils types; light-textured sands do not require as much time to wet up as finer-textured soils (loams and clays) but retain less water and have a lower water-holding capacity. Management factors such as cultivation and stubble retention will affect the amount of water the soil can hold. The greater the soil disturbance, the higher the loss of soil water.

In non-wetting or water-repellent soils, the accumulation of waxy organic matter in the soil surface can result in uneven wetting of the soil profile and poor emergence of crops.

Oxygen is essential to the germination process. In waterlogged soils, the soil oxygen is displaced by water, and the seeds can't germinate because of the very low levels of oxygen.

17.2 Moisture Stress Effects and Recovery Behavior of Important Crops

(a) **Water relations:** Moisture stress alters the water status by its influence on absorption, translocation, and transpiration. The lag in absorption behind transpiration results in loss of turgor as a result of increase in the atmospheric dryness.
(b) **Photosynthesis:** Photosynthesis is reduced by moisture stress due to reduction in photosynthetic rate, chlorophyll content, leaf area, and increase in assimilate saturation in leaves (due to lack of translocation).
(c) **Respiration:** Respiration increases with mild drought but more drought lowers water content and respiration.
(d) **Anatomical changes:** Decrease in size of the cells and intercellular spaces, thicker cell wall, greater development of mechanical tissue. Stomata per unit leaf tend to increase.
(e) **Metabolic reaction:** Almost all metabolic reactions are affected by water deficits.
(f) **Hormonal relationships:** The activity of growth-promoting hormones like cytokinin, gibberellic acid, and indole acetic acid decreases and growth-regulating hormone like abscisic acid, ethylene, etc., increases.
(g) **Nutrition:** The fixation, uptake, and assimilation of nitrogen is affected. Since dry matter production is considerably reduced, the uptake of NPK is reduced.
(h) **Growth and development:** Decrease in the growth of leaves, stems, and fruits. Maturity is delayed if drought occurs before flowering, while it advances if drought occurs after flowering.
(i) **Reproduction and grain growth:** Drought at flowering and grain development determines the number of fruits and individual grain weight, respectively. Panicle initiation in cereals is critical, while drought at anthesis may lead to drying of pollen. Drought at grain development reduces yield, while vegetative and grain filling stages are less sensitive to moisture stress.
(j) **Yield:** The effect on yield depends hugely on what proportion of the total dry matter is considered as useful material to be harvested. If it is aerial and underground parts, effect of drought is as sensitive as total growth. When the yield consists of seeds as in cereals, moisture stress at flowering is detrimental. When the yield is fiber or chemicals where economic product is a small fraction of total dry matter, moderate stress on growth does not have an adverse effect on yields.

17.3 Estimation of Moisture Index and Aridity Index

Moisture Index

A term based on the computation of an annual moisture budget by Thornthwaite and Mather (1955) and calculated from the aridity and humidity indices, as

$$I_m = 100 \times (S - D)/\mathrm{PE},$$

where I_m is the moisture index, S is the water surplus in months when precipitation exceeds evapotranspiration (ET), D is the water deficit in months when evapotranspiration exceeds precipitation, and PE is the potential evaporation

Moisture Index Formula

$$I_m = \text{humidity index} - 0.6(\text{aridity index})$$

$$I_m = \frac{100\ \mathrm{s} - 60d}{n}$$

Aridity Index

An aridity index (AI) is a numerical indicator of the degree of dryness of the climate at a given location. A number of aridity indices have been proposed; these indicators serve to identify, locate, or delimit regions that suffer from a deficit of available water, a condition that can severely affect the effective use of the land for such activities as agriculture or stock-farming.

In 1948, C. W. Thornthwaite proposed an AI defined as:

$$\mathrm{AI_T} = 100^* d/n$$

where the water deficiency d is calculated as the sum of the monthly differences between precipitation and potential evapotranspiration for those months when the normal precipitation is less than the normal evapotranspiration, and where n stands for the sum of monthly values of potential evapotranspiration for the deficient months (Huschke, 1959).

In the preparations leading to the UN Conference on Desertification (UNCOD), the United Nations Environment Programme (UNEP) issued a dryness map based on a different aridity index, proposed originally by Budyko (1958) and defined as follows:

$$\mathrm{AI_B} = 100^* R/LP$$

where R is the mean annual net radiation (also known as the net radiation balance), P is the mean annual precipitation, and L is the latent heat of vaporization for water.

More recently, the UNEP has adopted yet another index of aridity, defined as:

$$\mathrm{AI_U} = 100^* P/\mathrm{PET}$$

where PET is the potential evapotranspiration and P is the average annual precipitation (UNEP, 1992). Here also, PET and P must be expressed in the same units, e.g., in millimeters. In this latter case, the boundaries that define various degrees of aridity and the approximate areas involved are as follows:

Classification	Aridity index	Global land area
Hyperarid	AI < 0.05	7.5%
Arid	0.05 < AI < 0.20	12.1%
Semi-arid	0.20 < AI < 0.50	17.7%
Dry subhumid	0.50 < AI < 0.65	9.9%

17.4 Spray of Antitranspirants and Their Effect on Crops

Antitranspirants are any substances used to prevent water loss from transpiring plant surfaces. The term "antitranspirants" is also used interchangeably with the term "transpiration suppressants." The most effective antitranspirants can cut down transpiration losses by 30–40%. There are four different kinds of anti-perspirants.

(a) **Stomatal closing type:** Stomata on the leaf surface are the primary conduits for transpiration. In small quantities, several fungicides like PMA (phenyl mercuric acetate) and herbicides like atrazine act as antitranspirants by blocking stomata. Photosynthesis in mesophyll is known to be inhibited by PMA. Despite reports of success from research conducted in glasshouses, their utility in actual outdoor settings is limited.

(b) **Film-forming type:** Due to the creation of a physical barrier, the plastic and waxy components that form a thin layer on the surface of the leaf prevent water from escaping. Since these compounds also inhibit photosynthesis, their effectiveness is constrained. Antitranspirants of the film-forming type should have the following desirable qualities: they should create a thin layer, be more resistant to the passage of water vapor than carbon dioxide, and the film should maintain continuity and not tear. These antitranspirants that produce films can be either thin film or thick film.
Thin film-forming type: Hexadeconol
Thick film-forming type: Mobileaf, Polythene S-60

(c) **Leaf reflectant type:** These are the white substances that cover the leaves and raise their leaf reflectivity (albedo). By reflecting the radiation, they lower leaf temperatures and the gradient of the vapor pressure between the leaf and the atmosphere, which lowers transpiration. The leaf temperature is decreased by 34 °C, and transpiration is decreased by 22–28% with 5% of kaolin spray. Additionally utilized as reflectant anti-perspirants are celite and hydrated lime.

(d) **Growth retardant type:** These substances cause the roots to grow faster and the shoots to grow slower, allowing the plants to reduce their surface for transpiration and withstand drought conditions. The root/shoot ratio is increased.

For example: Phosphon–D, maleic hydrazide (MH), Cycocel—(2-chloroethyl) trimethyl ammonium chloride (CCC)

In general, photosynthesis is minimized by antitranspirants. Their application is therefore restricted to preventing crop death in situations of extreme moisture stress. The crop can make use of subsequent rainfall if survives. Antitranspirants can also help nursery plants recover more quickly from transplant shock. They are somewhat useful in horticulture crops and nurseries. Fruits are less likely to shrink after harvest when waxy materials are utilized.

Spray of Antitranspirants

Many landscapers apply their annual anti-desiccant sprays to broadleaf evergreen plants during the weeks of late fall and early winter. Evergreens will continue to transpire given certain conditions, although plants go dormant during the winter. During the dry winter months when the ground is frozen, these applications help reduce excessive water loss from leaves. An antitranspirant is a film-forming complex of polyethylenes and polyterpenes that when applied to foliage will reduce the moisture vapor transmission rate. A significant amount of water loss can also occur directly through the leaf cuticle or epidermis although much of the transpiration from leaves occur through small openings (stomates) under the leaves.

Plant moisture loss during the winter months is reduced between 15 and 20%, when 1 gallon of an antitranspirant is applied to 10–20 gallons of water. Moreover, the sprayed film produces a glossy sheet on broadleaf evergreens at the 5–10% dilution rate. Some labels indicate a second application during mid-winter provides best results, whereas other product labels state that a single application will last through the winter season. During late-winter periods, some of the most severe moisture loss from broadleaf evergreens typically occurs (e.g., February). In addition, on plants growing in full sun on warmer, windy days, the rate of transpiration will be greatest. Hence, an anti-desiccant in place during this time and under these conditions will usually be most beneficial.

Precautions

It is crucial to read the label included with all antitranspirant products all the time. Notice that some of these products should not be applied to evergreens having scale-like leaf foliage (i.e., junipers, arborvitae, *Cryptomeria*, *Chamaecyparis*, and Leyland cypress) and designed to be used only with needled evergreens and broadleaf. Winter-type injuries may be encouraged by using some anti-desiccants on scale leaf evergreens. In addition, when some antitranspirants are sprayed late in the afternoon, they dry more slowly during the evening hours, causing the film ingredients to separate out. This may cause a persistent frosty, white film that is not aesthetically pleasing to occur on the surface of broadleaf evergreens (e.g., *Rhododendrons*) appearing later in the winter. To reduce internal sprayer parts from becoming sticky and clogged with dried concentrate, sprayers should be washed out immediately after use with warm soapy water.

17.5 Water Use Efficiency

The ratio of water used in plant metabolism to water lost by the plant through transpiration is known as water use efficiency (WUE).

It is expressed in kg/ha cm. The following formula is used to calculate the proportion of water delivered and beneficially used on the project.

$$E_u = \frac{W_u}{W_d} \times 100$$

E_u = water use efficiency, %, W_u = beneficially used water, W_d = delivered water.

Water use efficiency is also defined as (i) crop water use efficiency and (ii) field water efficiency.

(a) Crop Water Use Efficiency: The ratio of yield of crop (Y) to the amount of water depleted by crop in evapotranspiration (ET) is known as crop water use efficiency.

$$\text{CWUE} = \frac{Y}{\text{ET}}$$

where CWUE = crop water use efficiency, Y = crop yield, ET = evapotranspiration

CWUE is otherwise called consumptive water use efficiency. It is the ratio of crop yield (Y) to the sum of the amount of water taken up and used for crop growth (G), evaporated directly from the soil surface (E) and transpired through foliage (T) or consumptive use (C_u).

$$\text{CWUE} = \frac{Y}{G + E + T}$$

where, $(G + E + T) = C_u$

In other words ET is C_u since water used for crop growth is negligible.

$$\text{CWUE} = \frac{Y}{\text{CU}}$$

It is expressed in kg/ha/mm or kg/ha/cm.

(b) Field Water Use Efficiency: The ratio of yield of crop (Y) to the total amount of water used in the field is known as field water efficiency

$$\text{FWUE} = \frac{Y}{\text{WR}}$$

where FWUE = field water use efficiency, WR = water requirement

This is the ratio of crop yield to the amount of water used in the field (WR) including growth (G), direct evaporation from the soil surface (E), transpiration (T), and deep percolation loss (D).

$$\text{FWUE} = \frac{Y}{G + E + T + D}$$

$$G + E + T + D = \text{WR}$$

It is expressed in kg/ha/mm (or) kg/ha/cm

References

Budyko MI (1958) The heat balance of the Earth's surface, trs. Nina A. Stepanova, US Department of Commerce, Washington, D.C., 259 p

Huschke RE (ed) (1959) Glossary of meteorology. American Meteorological Society, Boston, 638 pp

Thornthwaite CW, Mather JR (1955) The water balance. Publications in Climatology, Centerton

UNEP (1992) World atlas of desertification

GPSR Compliance

The European Union's (EU) General Product Safety Regulation (GPSR) is a set of rules that requires consumer products to be safe and our obligations to ensure this.

If you have any concerns about our products, you can contact us on ProductSafety@springernature.com

In case Publisher is established outside the EU, the EU authorized representative is:

Springer Nature Customer Service Center GmbH
Europaplatz 3
69115 Heidelberg, Germany

The manufacturer's authorised representative in the EU is Springer Nature Customer Service Centre GmbH, Europaplatz 3, 69115 Heidelberg, Germany. If you have any concerns regarding our products, please contact ProductSafety@springernature.com

Printed and bound by CPI Group (UK) Ltd, Croydon, CR0 4YY
30/09/2026
02225264-0012